Delmar's Test Preparation Series

Medium/Heavy Truck Test

Brakes (Test T4)

**Technical Advisor
John F. Kershaw**

Africa • Australia • Canada • Denmark • Japan • Mexico • New Zealand • Philippines
Puerto Rico • Singapore • Spain • United Kingdom • United States

NOTICE TO THE READER

Publisher does not warrant or guarantee any of the products described herein or perform any independent analysis in connection with any of the product information contained herein. Publisher does not assume, and expressly disclaims, any obligation to obtain and include information other than that provided to it by the manufacturer.

The reader is expressly warned to consider and adopt all safety precautions that might be indicated by the activities herein and to avoid all potential hazards. By following the instructions contained herein, the reader willingly assumes all risks in connection with such instructions.

The Publisher makes no representation or warranties of any kind, including but not limited to, the warranties of fitness for particular purpose or merchantability, nor are any such representations implied with respect to the material set forth herein, and the publisher takes no responsibility with respect to such material. The publisher shall not be liable for any special, consequential, or exemplary damages resulting, in whole or part, from the readers' use of, or reliance upon, this material.

Delmar Staff:
Business Unit Director: Alar Elken
Product Development Manager: Jack Erjavec
Executive Marketing Manager: Maura Theriault
Channel Manager: Mona Caron
Executive Production Manager: Mary Ellen Black
Cover Design: Paul Roseneck
Cover Image: © 1998 Corbis Corp.

COPYRIGHT © 2000
Delmar is a division of Thomson Learning. The Thomson Learning logo is a registered trademark used herein under license.

Printed in Canada
2 3 4 5 6 7 8 9 10 XXX 05 04 03 02 01 00 99

For more information, contact Delmar, 3 Columbia Circle, PO Box 15015, Albany, NY 12212-0515; or find us on the World Wide Web at http://www.delmar.com.

All rights reserved Thomson Learning 2000. The text of this publication, or any part thereof, may not be reproduced or transmitted in any form or by any means, electronics or mechanical, including photocopying, recording, storage in an information retrieval system, or otherwise, without prior permission of the publisher.

You can request permission to use material from this text through the following phone and fax numbers. Phone: 1-800-730-2214; Fax 1-800-730-2215; or visit our Web site at http://www.thomsonrights.com.

ISBN 0-7668-0562-X

Contents

Preface vi

Section 1 The History of ASE

History 1
    NIASE 1
    The Series and Individual Tests 2
    A Brief Chronology 2
    By the Numbers 3
    ASE 4

Section 2 Take and Pass Every ASE Test

ASE Testing 7
    Who Writes the Questions? 7
    Testing 8
    Be Test-Wise 8
    Before the Test 8
    Objective Tests 9
    Taking an Objective Test 9
    During the Test 10
    Review Your Answers 10
    Don't Be Distracted 10
    Use Your Time Wisely 11
    Don't Cheat 11
    Be Confident 11
    Anxiety and Fear 12
    Getting Rid of Fear 12
    Effective Study 13
    Make Study Definite 13
    The Urge to Learn 14
    Concentrate 14
        Get Sufficient Sleep 15
        Arrange Your Area 15
        Don't Daydream 15

Study Regularly 16
Keep a Record 17
Scoring the ASE Test 17
Understand the Test Results 18
    Brakes (Test T4) 18
"Average" 19
So, How Did You Do? 19

Section 3 Are You Sure You're Ready for Test T4?

Pretest 21
    Answers to the Test Questions for the Pretest 23
    Explanations to the Answers for the Pretest 23
Types of Questions 26
    Multiple-Choice Questions 26
    EXCEPT Questions 26
    Technician A, Technician B Questions 27
    Questions with a Figure 28
    Most-Likely Questions 28
    LEAST-Likely Questions 29
    Summary 29
Testing Time Length 30
    Monitor Your Progress 30
    Registration 30

Section 4 An Overview of the System

Brakes (Test T4) 31
  Task List and Overview 32
    A. Air Brakes Diagnosis and Repair (38 Questions) ... 32
        1. Air Supply and Service Systems (17 Questions) ... 32
        2. Mechanical/Foundation (11 Questions) 40
        3. Parking Brakes (5 Questions) 44
        4. Antilock Brake System (ABS) (5 Questions) ... 45
    B. Hydraulic Brakes Diagnosis and Repair (18 Questions) ... 47
        1. Hydraulic System (8 Questions) 47
        2. Mechanical System (6 Questions) 51
        3. Power Assist Units and Miscellaneous (4 Questions) ... 52
    C. Wheel Bearings Diagnosis and Repair (4 Questions) ... 53

Section 5 Sample Test for Practice

Sample Test 55

Section 6 Additional Test Questions for Practice

Additional Test Questions 69

Section 7 Appendices

Answers to the Test Questions for the Sample Test Section 5 91
Explanations to the Answers for the Sample Test Section 5 92
Answers to the Test Questions for the Additional Test Questions
    Section 6 102
Explanations to the Answers for the Additional Test Questions
    Section 6 103
Glossary 121

Preface

This book is just one of a comprehensive series designed to prepare technicians to take and pass every ASE test. Delmar's series covers all of the Automotive tests A1 through A8 as well as Advanced Engine Performance L1 and Parts Specialist P2. The series also covers the five Collision Repair tests and the eight Medium/Heavy Duty truck test.

Before any book in this series was written, Delmar staff met with and surveyed technicians and shop owners who have taken ASE tests and have used other preparatory materials. We found that they wanted, first and foremost, *lots* of practice tests and questions. Each book in our series contains a general knowledge pretest, a sample test, and additional practice questions. You will be hard-pressed to find a test prep book with more questions for you to practice with. We have worked hard to ensure that these questions match the ASE style in types of questions, quantities, and level of difficulty.

Technicians also told us that they wanted to understand the ASE test and to have practical information about what they should expect. We have provided that as well, including a history of ASE and a section devoted to helping the technician "Take and Pass Every ASE Test" with case studies, test-taking strategies, and test formats.

Finally, techs wanted refresher information and reference. Each of our books includes an overview section that is referenced to the task list. The complete task lists for each test appear in each book for the user's reference. There is also a complete glossary of terms for each booklet.

So whether you're looking for a sample test and a few extra questions to practice with or a complete introduction to ASE testing, with support for preparing thoroughly, this book series is an excellent answer.

We hope you benefit from this book and that you pass every ASE test you take!

Your comments, both positive and negative, are certainly encouraged! Please contact us at:

Automotive Editor
Delmar Publishers
3 Columbia Circle
Box 15015
Albany, NY 12212-5015

The History of ASE

History

Originally known as The National Institute for Automotive Service Excellence (NIASE), today's ASE was founded in 1972 as a nonprofit, independent entity dedicated to improving the quality of automotive service and repair through the voluntary testing and certification of automotive technicians. Until that time, consumers had no way of distinguishing between competent and incompetent automotive technicians. In the mid-1960s and early 1970s, efforts were made by several automotive industry affiliated associations to respond to this need. Though the associations were nonprofit, many regarded certification test fees merely as a means of raising additional operating capital. Also, some associations, having a vested interest, produced test scores heavily weighted in the favor of its members.

NIASE

From these efforts a new independent, nonprofit association, the National Institute for Automotive Service Excellence (NIASE), was established much to the credit of two educators, George R. Kinsler, Director of Program Development for the Wisconsin Board of Vocational and Adult Education in Madison, WI, and Myron H. Appel, Division Chairman at Cypress College in Cypress, CA.

Early efforts were to encourage voluntary certification in four general areas:

TEST AREA	TITLES
I. Engine	Engines, Engine Tune-Up, Block Assembly, Cooling and Lube Systems, Induction, Ignition, and Exhaust
II. Transmission	Manual Transmissions, Drive Line and Rear Axles, and Automatic Transmissions
III. Brakes and Suspension	Brakes, Steering, Suspension, and Wheels
IV. Electrical/Air Conditioning	Body/Chassis, Electrical Systems, Heating, and Air Conditioning

In early NIASE tests, Mechanic A, Mechanic B type questions were used. Over the years the trend has not changed, but in mid-1984 the term was changed to Technician A, Technician B to better emphasize sophistication of the skills needed to perform successfully in the modern motor vehicle industry. In certain tests the term used is Estimator A/B, Painter A/B, or Parts Specialist A/B. At about that same time, the logo was changed from "The Gear" to "The Blue Seal," and the organization adopted the acronym ASE for Automotive Service Excellence.

Since those early beginnings, several other related trades have been added. ASE now administers a comprehensive series of certification exams for automotive and light truck repair technicians, medium and heavy truck repair technicians, alternate fuels technicians, engine machinists, collision repair technicians, school bus repair technicians, and parts specialists.

The Series and Individual Tests

- Automotive and Light Truck Technician; consisting of: Engine Repair—Automatic Transmission/Transaxle—Manual Drive Train and Axles—Suspension and Steering—Brakes—Electrical/Electronic Systems—Heating and Air Conditioning—Engine Performance
- Medium and Heavy Truck Technician; consisting of: Gasoline Engines—Diesel Engines—Drive Train—Brakes—Suspension and Steering—Electrical/Electronic Systems—HVAC—Preventive Maintenance Inspection
- Alternate Fuels Technician; consisting of: Compressed Natural Gas Light Vehicles
- Advanced Series; consisting of: Automobile Advanced Engine Performance and Advanced Diesel Engine Electronic Diesel Engine Specialty
- Collision Repair Technician; consisting of: Painting and Refinishing—Non-Structural Analysis and Damage Repair—Structural Analysis and Damage Repair—Mechanical and Electrical Components—Damage Analysis and Estimating
- Engine Machinist Technician; consisting of: Cylinder Head Specialist—Cylinder Block Specialist—Assembly Specialist
- School Bus Repair Technician; consisting of: Body Systems and Special Equipment—Drive Train—Brakes—Suspension and Steering—Electrical/Electronic Systems—Heating and Air Conditioning
- Parts Specialist; consisting of: Automobile Parts Specialist—Medium/Heavy Truck Parts Specialist

A Brief Chronology

1970–1971   Original questions were prepared by a group of forty auto mechanics teachers from public secondary schools, technical institutes, community colleges, and private vocational schools. These questions were then professionally edited by testing specialists at Educational Testing Service (ETS) at Princeton, New Jersey, and thoroughly reviewed by training specialists associated with domestic and import automotive companies.

1971   July: About eight hundred mechanics tried out the original test questions at experimental test administrations.

1972   November and December: Initial NIASE tests administered at 163 test centers. The original automotive test series consisted of four tests containing eighty questions each. Three hours were allotted for each test. Those who passed all four tests were designated Certified General Auto Mechanic (GAM).

1973   April and May: Test 4 was increased to 120 questions. Time was extended to four hours for this test. There were now 182 test centers. Shoulder patch insignias were made available.

The History of ASE

	November: Automotive series expanded to five tests. Heavy-Duty Truck series of six tests introduced.
1974	November: Automatic Transmission (Light Repair) test modified. Name changed to Automatic Transmission.
1975	May: Collision Repair series of two tests is introduced.
1978	May: Automotive recertification testing is introduced.
1979	May: Heavy-Duty Truck recertification testing is introduced.
1980	May: Collision Repair recertification testing is introduced.
1982	May: Test administration providers switched from Educational Testing Service (ETS) to American College Testing (ACT). Name of Automobile Engine Tune-Up test changed to Engine Performance test.
1984	May: New logo was introduced. ASE's "The Blue Seal" replaced NIASE's "The Gear." All reference to Mechanic A, Mechanic B was changed to Technician A, Technician B.
1990	November: The first of the Engine Machinist test series was introduced.
1991	May: The second test of the Engine Machinist test series was introduced. November: The third and final Engine Machinist test was introduced.
1992	May: Name of Heavy-Duty Truck Test series changed to Medium/Heavy Truck test series.
1993	May: Automotive Parts Specialist test introduced. Collision Repair expanded to six tests. Light Vehicle Compressed Natural Gas test introduced. Limited testing begins in English-speaking provinces of Canada.
1994	May: Advanced Engine Performance Specialist test introduced.
1996	May: First three tests for School Bus Technician test series introduced. November: A Collision Repair test is added.
1997	May: A Medium/Heavy Truck test is added.
1998	May: A diesel advanced engine test is introduced: Electronic Diesel Engine Diagnosis Specialist. A test is added to the School Bus test series.

By the Numbers

Following are the approximate number of ASE technicians currently certified by category. The numbers may vary from time to time but are reasonably accurate for any given period. More accurate data may be obtained from ASE, which provides updates twice each year, in May and November after the Spring and Fall test series.

There are more than 338,000 Automotive Technicians with over 87,000 at Master Technician (MA) status. There are 47,000 Truck Technicians with over 19,000 at Master Technician (MT) status. There are 46,000 Collision Repair/Refinish Technicians with 7,300 at Master Technician (MB) status. There are 1,200 Estimators. There are 6,700 Engine Machinists with over 2,800 at Master Machinist Technician (MM) status. There are also 28,500 Automobile Advanced Engine Performance Technicians and over 2,700 School Bus Technicians for a combined total of more than 403,000 Repair Technicians. To this number, add over 22,000 Automobile Parts Specialists, and over 2,000 Truck Parts Specialists for a combined total of over 24,000 parts specialists.

There are over 6,400 ASE Technicians holding both Master Automotive Technician and Master Truck Technician status, of which 350 also hold Master Body Repair status. Almost 200 of these Master Technicians also hold Master Machinist status and five Technicians are certified in all ASE specialty areas.

Almost half of ASE certified technicians work in new vehicle dealerships (45.3 percent). The next greatest number work in independent garages with 19.8 percent. Next is tire dealerships with 9 percent, service stations at 6.3 percent, fleet shops at 5.7 percent, franchised volume retailers at 5.4 percent, paint and body shops at 4.3 percent, and specialty shops at 3.9 percent.

Of over 400,000 automotive technicians on ASE's certification rosters, almost 2,000 are female. The number of female technicians is increasing at a rate of about 20 percent each year. Women's increasing interest in the automotive industry is further evidenced by the fact that, according to the National Automobile Dealers Association (NADA), they influence 80 percent of the decisions of the purchase of a new automobile and represent 50 percent of all new car purchasers. Also, it is interesting to note that 65 percent of all repair and maintenance service customers are female.

The typical ASE certified technician is 36.5 years of age, is computer literate, deciphers a half-million pages of technical manuals, spends one hundred hours per year in training, holds four ASE certificates, and spends about $27,000 for tools and equipment. Twenty-seven percent of today's skilled ASE certified technicians attended college, many having earned an Associate of Science degree in Automotive Technology.

ASE

ASE's mission is to improve the quality of vehicle repair and service in the United States through the testing and certification of automotive repair technicians. Prospective candidates register for and take one or more of ASE's thirty-three exams. The tests are grouped into specialties for automobile, medium/heavy truck, school bus, and collision repair technicians as well as engine machinists, alternate fuels technicians, and parts specialists.

Upon passing at least one exam and providing proof of two years of related work experience, the technician becomes ASE certified. A technician who passes a series of exams earns ASE Master Technician status. An automobile technician, for example, must pass eight exams for this recognition.

The tests, conducted twice a year at over seven hundred locations around the country, are administered by American College Testing (ACT). They stress real-world diagnostic and repair problems. Though a good knowledge of theory is helpful to the technician in answering many of the questions, there are no questions specifically on theory. Certification is valid for five years. To retain certification, the technician must be retested to renew his or her certificate.

The automotive consumer benefits because ASE certification is a valuable yardstick by which to measure the knowledge and skills of individual technicians, as well as their commitment to their chosen profession. It is also a tribute to the repair facility employing ASE certified technicians. ASE certified technicians are permitted to wear blue and white ASE shoulder insignia, referred to as the "Blue Seal of Excellence," and carry credentials listing their areas of expertise. Often employers display their technicians' credentials in the customer waiting area. Customers look for facilities that display ASE's Blue Seal of Excellence logo on outdoor signs, in the customer waiting area, in the telephone book (Yellow Pages), and in newspaper advertisements.

The tests stress repair knowledge and skill. All test takers are issued a score report. In order to earn ASE certification, a technician must pass one or more of the exams and present proof of two years of relevant hands-on work experience. ASE certifications are valid for five years, after which time technicians must retest in order to keep up with changing technology and to remain in the ASE program. A nominal registration and test fee is charged.

The History of ASE

To become part of the team that wears ASE's Blue Seal of Excellence®, please contact:

National Institute for Automotive Service Excellence
13505 Dulles Technology Drive
Herndon, VA 20171-3421

2 Take and Pass Every ASE Test

ASE Testing

Participating in an Automotive Service Excellence (ASE) voluntary certification program gives you a chance to show your customers that you have the "know-how" needed to work on today's modern vehicles. The ASE certification tests allow you to compare your skills and knowledge to the automotive service industry's standards for each specialty area.

If you are the "average" automotive technician taking this test, you are in your mid-thirties and have not attended school for about fifteen years. That means you probably have not taken a test in many years. Some of you, on the other hand, have attended college or taken postsecondary education courses and may be more familiar with taking tests and with test-taking strategies. There is, however, a difference in the ASE test you are preparing to take and the educational tests you may be accustomed to.

Who Writes the Questions?

The questions on an educational test are generally written, administered, and graded by an educator who may have little or no practical hands-on experience in the test area. The questions on all ASE tests are written by service industry experts familiar with all aspects of the subject area. ASE questions are entirely job-related and designed to test the skills that you need to know on the job.

The questions originate in an ASE "item-writing" workshop where service representatives from domestic and import automobile manufacturers, parts and equipment manufacturers, and vocational educators meet in a workshop setting to share their ideas and translate them into test questions. Each test question written by these experts is reviewed by all of the members of the group. The questions deal with the practical problems of diagnosis and repair that are experienced by technicians in their day-to-day hands-on work experiences.

All of the questions are pretested and quality-checked in a nonscoring section of tests by a national sample of certifying technicians. The questions that meet ASE's high standards of accuracy and quality are then included in the scoring sections of future tests. Those questions that do not pass ASE's stringent tests are sent back to the workshop or are discarded. ASE's tests are monitored by an independent proctor and are administered and machine-scored by an independent provider, American College Testing (ACT). All ASE tests have a three-year revision cycle.

Testing

If you think about it, we are actually tested on about everything we do. As infants, we were tested to see when we could turn over and crawl, later when we could walk or talk. As adolescents, we were tested to determine how well we learned the material presented in school and in how we demonstrated our accomplishments on the athletic field. As working adults, we are tested by our supervisors on how well we have completed an assignment or project. As nonworking adults, we are tested by our family on everyday activities, such as housekeeping or preparing a meal. Testing, then, is one of those facts of life that begins in the cradle and follows us to the grave.

Testing is an important fact of life that helps us to determine how well we have learned our trade. Also, tests often help us to determine what opportunities will be available to us in the future. To become ASE certified, we are required to take a test in every subject in which we wish to be recognized.

Be Test-Wise

In spite of the widespread use of tests, most technicians are not very test-wise. An ability to take tests and score well is a skill that must be acquired. Without this knowledge, the most intelligent and prepared technician may not do well on a test.

We will discuss some of the basic procedures necessary to follow in order to become a test-wise technician. Assume, if you will, that you have done the necessary study and preparation to score well on the ASE test.

Different approaches should be used for taking different types of tests. The different basic types of tests include: essay, objective, multiple-choice, fill-in-the-blank, true-false, problem solving, and open book. All ASE tests are of the four-part multiple-choice type.

Before discussing the multiple-choice type test questions, however, there are a few basic principles that should be followed before taking any test.

Before the Test

Do not arrive late. Always arrive well before your test is scheduled to begin. Allow ample time for the unexpected, such as traffic problems, so you will arrive on time and avoid the unnecessary anxiety of being late.

Always be certain to have plenty of supplies with you. For an ASE test, three or four sharpened soft lead (#2) pencils, a pocket pencil sharpener, erasers, and a watch are all that are required.

Do not listen to pretest chatter. When you arrive early, you may hear other technicians testing each other on various topics or making their best guess as to the probable test questions. At this time, it is too late to add to your knowledge. Also the rhetoric may only confuse you. If you find it bothersome, take a walk outside the test room to relax and loosen up.

Read and listen to all instructions. It is important to read and listen to the instructions. Make certain that you know what is expected of you. Listen carefully to verbal instructions and pay particular attention to any written instructions on the test paper. Do not dive into answering questions only to find out that you have answered the wrong question by not following instructions carefully. It is difficult to make a high score on a test if you answer the wrong questions.

These basic principles have been violated in most every test ever given. Try to remember them. They are essential for success.

Objective Tests

A test is called an objective test if the same standards and conditions apply to everyone taking the test and there is only one correct answer to each question. Objective tests primarily measure your ability to recall information. A well-designed objective test can also test your ability to understand, analyze, interpret, and apply your knowledge. Objective tests include true-false, multiple-choice, fill-in-the-blank, and matching questions.

Objective questions, not generally encountered in a classroom setting, are frequently used in standardized examinations. Objective tests are easy to grade and also reduce the amount of paperwork necessary to administer. The objective tests are used in entry-level programs or when very large numbers are being tested. ASE's tests consist exclusively of four-part multiple-choice objective questions in all of their tests.

Taking an Objective Test

The principles of taking an objective test are somewhat different from those used in other types of tests. You should first quickly look over the test to determine the number of questions, but do not try to read through all of the questions. In an ASE test, there are usually between forty and eighty questions, depending on the subject matter. Read through each question before marking your answer. Answer the questions in the order they appear on the test. Leave the questions blank that you are not sure of and move on to the next question. You can return to those unanswered questions after you have finished the others. They may be easier to answer at a later time after your mind has had additional time to consider them on a subconscious level. In addition, you might find information in other questions that will help you to answer some of them.

Do not be obsessed by the apparent pattern of responses. For example, do not be influenced by a pattern like **d**, **c**, **b**, **a**, **d**, **c**, **b**, **a** on an ASE test.

There is also a lot of folk wisdom about taking objective tests. For example, there are those who would advise you to avoid response options that use certain words such as *all*, *none*, *always*, *never*, *must*, and *only*, to name a few. This, they claim, is because nothing in life is exclusive. They would advise you to choose response options that use words that allow for some exception, such as *sometimes*, *frequently*, *rarely*, *often*, *usually*, *seldom*, and *normally*. They would also advise you to avoid the first and last option (A and D) because test writers, they feel, are more comfortable if they put the correct answer in the middle (B and C) of the choices. Another recommendation often offered is to select the option that is either shorter or longer than the other three choices because it is more likely to be correct. Some would advise you to never change an answer since your first intuition is usually correct.

Although there may be a grain of truth in this folk wisdom, ASE test writers try to avoid them and so should you. There are just as many **A** answers as there are **B** answers, just as many **D** answers as **C** answers. As a matter of fact, ASE tries to balance the answers at about 25 percent per choice **A**, **B**, **C**, and **D**. There is no intention to use "tricky" words, such as outlined above. Put no credence in the opposing words "sometimes" and "never," for example. When used in an ASE type question, one or both may be correct; one or both may be incorrect.

There are some special principles to observe on multiple-choice tests. These tests are sometimes challenging because there are often several choices that may seem possible, and it may be difficult to decide on the correct choice. The best strategy, in this case, is to first determine the correct answer before looking at the options. If you see the answer you decided on, you should still examine the options to make sure that none seem more correct than yours. If you do not know or are not sure of the answer, read each option very carefully and try to eliminate those options that you know to be wrong. That way, you can often arrive at the correct choice through a process of elimination.

If you have gone through all of the test and you still do not know the answer to some of the questions, then guess. Yes, guess. You then have at least a 25 percent chance of being correct. If you leave the question blank, you have no chance. In ASE tests, there is no penalty for being wrong. As the late President Franklin D. Roosevelt once advised a group of students, "It is common sense to take a method and try it. If it fails, admit it frankly and try another. But above all, try something."

During the Test

Mark your bubble sheet clearly and accurately. One of the biggest problems an adult faces in test-taking, it seems, is in placing an answer in the correct spot on a bubble sheet. Make certain that you mark your answer for, say, question 21, in the space on the bubble sheet designated for the answer for question 21. A correct response in the wrong bubble will probably be wrong. Remember, the answer sheet is machine scored and can only "read" what you have bubbled in. Also, do not bubble in two answers for the same question. For example, if you feel the answer to a particular question is **A** but think it may be **C**, do not bubble in both choices. Even if either **A** or **C** is correct, a double answer will score as an incorrect answer. It's better to take a chance with your best guess.

Review Your Answers

If you finish answering all of the questions on a test ahead of time, go back and review the answers of those questions that you were not sure of. You can often catch careless errors by using the remaining time to review your answers.

Don't Be Distracted

At practically every test, some technicians will invariably finish ahead of time and turn their papers in long before the final call. Do not let them distract or intimidate you. Either they knew too little and could not finish the test, or they were very self-confident and thought they knew it all. Perhaps they were trying to impress the proctor or other technicians about how much they know. Often you may hear them later talking about the information they knew all the while but forgot to respond on their answer sheet.

Use Your Time Wisely

It is not wise to use less than the total amount of time that you are allotted for a test. If there are any doubts, take the time for review. Any product can usually be made better with some additional effort. A test is no exception. It is not necessary to turn in your test paper until you are told to do so.

Don't Cheat

Some technicians may try to use a "crib sheet" during a test. Others may attempt to read answers from another technician's paper. If you do that, you are unquestionably assuming that someone else has a correct answer. You probably know as much, maybe more, than anyone else in the test room. Trust yourself. If you're still not convinced, think of the consequences of being caught. Cheating is foolish. If you are caught, you have failed the test.

Be Confident

The first and foremost principle in taking a test is that you need to know what you are doing, to be test-wise. It will now be presumed that you are a test-wise technician and are now ready for some of the more obscure aspects of test-taking.

An ASE-style test requires that you use the information and knowledge at your command to solve a problem. This generally requires a combination of information similar to the way you approach problems in the real world. Most problems, it seems, typically do not fall into neat textbook cases. New problems are often difficult to handle, whether they are encountered inside or outside the test room.

An ASE test also requires that you apply methods taught in class as well as those learned on the job to solve problems. These methods are akin to a well-equipped tool box in the hands of a skilled technician. You have to know what tools to use in a particular situation, and you must also know how to use them. In an ASE test, you will need to be able to demonstrate that you are familiar with and know how to use the tools.

You should begin a test with a completely open mind. At times, however, you may have to move out of your normal way of thinking and be creative to arrive at a correct answer. If you have diligently studied for at least one week prior to the test, you have bombarded your mind with a wide assortment of information. Your mind will be working with this information on a subconscious level, exploring the interrelationships among various facts, principles, and ideas. This prior preparation should put you in a creative mood for the test.

In order to reach your full potential, you should begin a test with the proper mental attitude and a high degree of self-confidence. You should think of a test as an opportunity to document how much you know about the various tasks in your chosen profession. If you have been diligently studying the subject matter, you will be able to take your test in serenity because your mind will be well organized. If you are confident, you are more likely to do well because you have the proper mental attitude. If, on the other hand, your confidence is low, you are bound to do poorly. It is a self-fulfilling prophecy.

Perhaps you have heard athletic coaches talk about the importance of confidence when competing in sports. Mental confidence helps an athlete to perform at the highest level and gain an advantage over competitors. Taking a test is much like an athletic event. You are competing against yourself, in a certain sense, because you will be trying to approach perfection in determining your answers. As in any competition, you should aim your sights high and be confident that you can reach the apex.

Anxiety and Fear

Many technicians experience anxiety and fear at the very thought of taking a test. Many worry, become nervous, and even become ill at test time because of the fear of failure. Many often worry about the criticism and ridicule that may come from their employer, relatives, and peers. Some worry about taking a test because they feel that the stakes are very high. Those who spent a great amount of time studying may feel they must get a high grade to justify their efforts. The thought of not doing well can result in unnecessary worry. They become so worried, in fact, that their reasoning and thinking ability is impaired, actually bringing about the problem they wanted to avoid.

The fear of failure should not be confused with the desire for success. It is natural to become "psyched-up" for a test in contemplation of what is to come. A little emotion can provide a healthy flow of adrenaline to peak your senses and hone your mental ability. This improves your performance on the test and is a very different reaction from fear.

Most technician's fears and insecurities experienced before a test are due to a lack of self-confidence. Those who have not scored well on previous tests or have no confidence in their preparation are those most likely to fail. Be confident that you will do well on your test and your fears should vanish. You will know that you have done everything possible to realize your potential.

Getting Rid of Fear

If you have previously experienced fear of taking a test, it may be difficult to change your attitude immediately. It may be easier to cope with fear if you have a better understanding of what the test is about. A test is merely an assessment of how much the technician knows about a particular task area. Tests, then, are much less threatening when thought of in this manner. This does not mean, however, that you should lower your self-esteem simply because you performed poorly on a test.

You can consider the test essentially as a learning device, providing you with valuable information to evaluate your performance and knowledge. Recognize that no one is perfect. All humans make mistakes. The idea, then, is to make mistakes before the test, learn from them, and avoid repeating them on the test. Fortunately, this is not as difficult as it seems. Practical questions in this study guide include the correct answers to consider if you have made mistakes on the practice test. You should learn where you went wrong so you will not repeat them in the ASE test. If you learn from your mistakes, the stage is set for future growth.

If you understood everything presented up until now, you have the knowledge to become a test-wise technician, but more is required. To be a test-wise technician, you not only have to practice these principles, you have to diligently study in your task area.

Effective Study

The fundamental and vital requirement to induce effective study is a genuine and intense desire to achieve. This is more basic than any rule or technique that will be given here. The key requirement, then, is a driving motivation to learn and to achieve.

If you wish to study effectively, first develop a desire to master your studies and sincerely believe that you will master them. Everything else is secondary to such a desire.

First, build up definite ambitions and ideals toward which your studies can lead. Picture the satisfaction of success. The attitude of the technician may be transformed from merely getting by to an earnest and energetic effort. The best direct stimulus to change may involve nothing more than the deliberate planning of your time. Plan time to study.

Another drive that creates positive study is an interest in the subject studied. As an automotive technician, you can develop an interest in studying particular subjects if you follow these four rules:

1. Acquire information from a variety of sources. The greater your interest in a subject, the easier it is to learn about it. Visit your local library and seek books on the subject you are studying. When you find something new or of interest, make inexpensive photocopies for future study.
2. Merge new information with your previous knowledge. Discover the relationship of new facts to old known facts. Modern developments in automotive technology take on new interest when they are seen in relation to present knowledge.
3. Make new information personal. Relate the new information to matters that are of concern to you. The information you are now reading, for example, has interest to you as you think about how it can help.
4. Use your new knowledge. Raise questions about the points made by the book. Try to anticipate what the next steps and conclusions will be. Discuss this new knowledge, particularly the difficult and questionable points, with your peers.

You will find that when you study with eager interest, you will discover it is no longer work. It is pleasure and you will be fascinated in what you study. Studying can be like reading a novel or seeing a movie that overcomes distractions and requires no effort or willpower. You will discover that the positive relationship between interest and effort works both ways. Even though you perhaps began your studies with little or no interest, simply staying with it helped you to develop an interest in your studies.

Obviously, certain subject matter studies are bound to be of little or no interest, particularly in the beginning. Parts of certain studies may continue to be uninteresting. An honest effort to master those subjects, however, nearly always brings about some level of interest. If you appreciate the necessity and reward of effective studying, you will rarely be disappointed. Here are a few important hints for gaining the determination that is essential to carrying good conclusions into actual practice.

Make Study Definite

Decide what is to be studied and when it is to be studied. If the unit is discouragingly long, break it into two or more parts. Determine exactly what is involved in the first part and learn that. Only then should you proceed to the next part. Stick to a schedule.

The Urge to Learn

Make clear to yourself the relation of your present knowledge to your study materials. Determine the relevance with regard to your long-range goals and ambitions.

Turn your attention away from real or imagined difficulties as well as other things that you would rather be doing. Some major distractions are thoughts of other duties and of disturbing problems. These distractions can usually be put aside, simply shunted off by listing them in a notebook. Most technicians have found that by writing interfering thoughts down, their minds are freed from annoying tensions.

Adopt the most reasonable solution you can find or seek objective help from someone else for personal problems. Personal problems and worry are often causes of ineffective study. Sometimes there are no satisfactory solutions. Some manage to avoid the problems or to meet them without great worry. For those who may wish to find better ways of meeting their personal problems, the following suggestions are offered:

1. Determine as objectively and as definitely as possible where the problem lies. What changes are needed to remove the problem, and which changes, if any, can be made? Sometimes it is wiser to alter your goals than external conditions. If there is no perfect solution, explore the others. Some solutions may be better than others.

2. Seek an understanding confidant who may be able to help analyze and meet your problems. Very often, talking over your problems with someone in whom you have confidence and trust will help you to arrive at a solution.

3. Do not betray yourself by trying to evade the problem or by pretending that it has been solved. If social problem distractions prevent you from studying or doing satisfactory work, it is better to admit this to yourself. You can then decide what can be done about it.

Once you are free of interferences and irritations, it is much easier to stay focused on your studies.

Concentrate

To study effectively, you must concentrate. Your ability to concentrate is governed, to a great extent, by your surroundings as well as your physical condition. When absorbed in study, you must be oblivious to everything else around you. As you learn to concentrate and study, you must also learn to overcome all distractions. There are three kinds of distractions you may face:

1. Distractions in the surrounding area, such as motion, noise, and the glare of lights. The sun shining through a window on your study area, for example, can be very distracting.

   Some technicians find that, for effective study, it is necessary to eliminate visual distractions as well as noises. Others find that they are able to tolerate moderate levels of auditory or visual distraction.

   Make sure your study area is properly lighted and ventilated. The lighting should be adequate but should not shine directly into your eyes or be visible out of the corner of your eye. Also, try to avoid a reflection of the lighting on the pages of your book.

   Whether heated or cooled, the environment should be at a comfortable level. For most, this means a temperature of 78°F–80°F (25.6°C–26.7°C) with a relative humidity of 45 to 50 percent.

2. Distractions arising from your body, such as a headache, fatigue, and hunger. Be in good physical condition. Eat wholesome meals at regular times. Try to eat with your family or friends whenever possible. Mealtime should be your recreational period. Do not eat a heavy meal for lunch, and do not resume studies immediately after eating lunch. Just after lunch, try to get some regular exercise, relaxation, and recreation. A little exercise on a regular basis is much more valuable than a lot of exercise only on occasion.
3. Distractions of irrelevant ideas, such as how to repair the garden gate, when you are studying for an automotive-related test.

The problems associated with study are no small matter. These problems of distractions are generally best dealt with by a process of elimination. A few important rules for eliminating distractions follow.

Get Sufficient Sleep

You must get plenty of rest even if it means dropping certain outside activities. Avoid cutting in on your sleep time; you will be rewarded in the long run. If you experience difficulty going to sleep, do something to take your mind off your work and try to relax before going to bed. Some suggestions that may help include a little reading, a warm bath, a short walk, a conversation with a friend, or writing that overdue letter to a distant relative. If sleeplessness is an ongoing problem, consult a physician. Do not try any of the sleep remedies on the market, particularly if you are on medication, without approval of your physician.

If you still have difficulty studying, a final rule may help. Sit down in a favorable place for studying, open your books, and take out your pencil and paper. In a word, go through the motions.

Arrange Your Area

Arrange your chair and work area. To avoid strain and fatigue, whenever possible, shift your position occasionally. Try to be comfortable; however, avoid being too comfortable. It is nearly impossible to study rigorously when settled back in a large easy chair or reclining leisurely on a sofa.

When studying, it is essential to have a plan of action, a time to work, a time to study, and a time for pleasure. If you schedule your day and adhere to the schedule, you will eliminate most of your efforts and worries. A plan that is followed, then, soon becomes the easy and natural routine of the day. Most technicians find it useful to have a definite place and time to study. A particular table and chair should always be used for study and intellectual work. This place will then come to mean study. To be seated in that particular location at a regularly scheduled time will automatically lead you to assume a readiness for study.

Don't Daydream

Daydreaming or mind-wandering is an enemy of effective study. Daydreaming is frequently due to an inadequate understanding of words. Use the Glossary or a dictionary to look up the troublesome word. Another frequent cause of daydreaming is a deficient background in the present subject matter. When this is the problem, go back and review the subject matter to obtain the necessary foundation. Just one hour of concentrated study is equivalent to ten hours with frequent lapses of daydreaming. Be on guard against mind-wandering, and pull yourself back into focus on every occasion.

Study Regularly

A system of regularity in study is believed by many scholars to be the secret of success. The daily time schedule must, however, be determined on an individual basis. You must decide how many hours of each day you can devote to your studies. Few technicians really are aware of where their leisure time is spent. An accurate account of how your days are presently being spent is an important first step toward creating an effective daily schedule.

	\multicolumn{7}{c}{Weekly Schedule}					
	Sun	Mon	Tues	Wed	Thu	Fri	Sat
6:00							
6:30							
7:00							
7:30							
8:00							
8:30							
9:00							
9:30							
10:00							
10:30							
11:00							
11:30							
NOON							
12:30							
1:00							
1:30							
2:00							
2:30							
3:00							
3:30							
4:00							
4:30							
5:00							
5:30							
6:00							
6:30							
7:00							
7:30							
8:00							
8:30							
9:00							
9:30							
10:00							
10:30							
11:00							
11:30							

The convenient form is for keeping an hourly record of your week's activities. If you fill in the schedule each evening before bedtime, you will soon gain some interesting and useful facts about yourself and your use of your time. If you think over the causes of wasted time, you can determine how you might better spend your time. A practical schedule can be set up by using the following steps.

1. Mark your fixed commitments, such as work, on your schedule. Be sure to include classes and clubs. Do you have sufficient time left? You can arrive at an estimate of the time you need for studying by counting the hours used during the present week. An often-used formula, if you are taking classes, is to multiply the number of hours you spend in class by two. This provides time for class studies. This is then added to your work hours. Do not forget time allocation for travel.

2. Fill in your schedule for meals and studying. Use as much time as you have available during the normal workday hours. Do not plan, for example, to do all of your studying between 11:00 P.M. and 1:00 A.M. Try to select a time for study that you can use every day without interruption. You may have to use two or perhaps three different study periods during the day.

3. List the things you need to do within a time period. A one-week time frame seems to work well for most technicians. The question you may ask yourself is: "What do I need to do to be able to walk into the test next week, or next month, prepared to pass?"

4. Break down each task into smaller tasks. The amount of time given to each area must also be settled. In what order will you tackle your schedule? It is best to plan the approximate time for your assignments and the order in which you will do them. In this way, you can avoid the difficulties of not knowing what to do first and of worrying about the other things you should be doing.

5. List your tasks in the empty spaces on your schedule. Keep some free time unscheduled so you can deal with any unexpected events, such as a dental appointment. You will then have a tentative schedule for the following week. It should be flexible enough to allow some units to be rearranged if necessary. Your schedule should allow time off from your studies. Some use the promise of a planned recreational period as a reward for motivating faithfulness to a schedule. You will more likely lose control of your schedule if it is packed too tightly.

Keep a Record

Keep a record of what you actually do. Use the knowledge you gain by keeping a record of what you are actually doing so you can create or modify a schedule for the following week. Be sure to give yourself credit for movement toward your goals and objectives. If you find that you cannot study productively at a particular hour, modify your schedule so as to correct that problem.

Scoring the ASE Test

You can gain a better perspective about tests if you know and understand how they are scored. ASE's tests are scored by American College Testing (ACT), a nonpartial, nonbiased organization having no vested interest in ASE or in the automotive industry. Each question carries the same weight as any other question. For example, if there are fifty questions, each is worth 2 percent of the total score. The passing grade is 70 percent. That means you must correctly answer thirty-five of the fifty questions to pass the test.

Understand the Test Results

The test results can tell you:
- where your knowledge equals or exceeds that needed for competent performance, or
- where you might need more preparation.

The test results *cannot* tell you:
- how you compare with other technicians, or
- how many questions you answered correctly.

Your ASE test score report will show the number of correct answers you got in each of the content areas. These numbers provide information about your performance in each area of the test. However, because there may be a different number of questions in each area of the test, a high percentage of correct answers in an area with few questions may not offset a low percentage in an area with many questions.

It may be noted that one does not "fail" an ASE test. The technician that does not pass is simply told "More Preparation Needed." Though large differences in percentages may indicate problem areas, it is important to consider how many questions were asked in each area. Since each test evaluates all phases of the work involved in a service specialty, you should be prepared in each area. A low score in one area could keep you from passing an entire test.

Note that a typical test will contain the number of questions indicated above each content area's description. For example:

Brakes (Test T4)

Content Area	Questions	Percent of Test
A. Air Brakes Diagnosis and Repair	38	63%
1. Air Supply and Service Systems (17)		
2. Mechanical/Foundation (11)		
3. Parking Brakes (5)		
4. Antilock Brake System (ABS) (5)		
B. Hydraulic Brakes Diagnosis and Repair	18	30%
1. Hydraulic System (8)		
2. Mechanical System (6)		
3. Power Assist Units and Miscellaneous (4)		
C. Wheel Bearings Diagnosis and Repair	4	7%
Total	*60	100%

Note: *The test could contain up to ten additional questions that are included for statistical research purposes only. Your answers to these questions will not affect your score, but since you do not know which ones they are, you should answer all questions in the test. The five-year Recertification Test will cover the same content areas as those listed above. However, the number of questions in each content area of the Recertification Test will be reduced by about one-half.*

"Average"

There is no such thing as average. You cannot determine your overall test score by adding the percentages given for each task area and dividing by the number of areas. It doesn't work that way because there generally are not the same number of questions in each task area. A task area with twenty questions, for example, counts more toward your total score than a task area with ten questions.

So, How Did You Do?

Your test report should give you a good picture of your results and a better understanding of your task areas of strength and weakness.

If you fail to pass the test, you may take it again at any time it is scheduled to be administered. You are the only one who will receive your test score. Test scores will not be given over the telephone by ASE nor will they be released to anyone without your written permission.

Are You Sure You're Ready for Test T4?

Pretest

The purpose of this pretest is to determine the amount of review that you may require prior to taking the Automative Service Excellence (ASE) medium/heavy truck test: Brakes (Test T4). If you answer all of the pretest questions correctly, complete the sample test in section 5 along with the additional test questions in section 6.

If two or more of your answers to the pretest questions are wrong, study section 4: An Overview of the System before continuing with the sample test and additional test questions.

The pretest answers and explanations are located at the end of the pretest.

1. The compressor for a truck air braking system is driven by:
   A. a separate electrical motor.
   B. the vehicle's engine.
   C. the airflow moving past its vanes.
   D. heated air from the engine.

2. What is another name for the supply tank?
   A. Reservoir tank
   B. Wet tank
   C. Dry tank
   D. Primary reservoir

3. A truck equipped with air brakes has parking brake release problems. Technician A says the cause could be a defective spring brake chamber. Technician B says the parking brake cable may be seized. Who is right?
   A. A only
   B. B only
   C. Both A and B
   D. Neither A nor B

4. Technician A says that trucks over 10,000 Gross Vehicle Weight Rating (GVWR) equipped with hydraulic brakes must use a dual-circuit hydraulic system. Technician B says that trucks over 10,000 GVWR are not required to meet any specific federal hydraulic requirements. Who is right?
   A. A only
   B. B only
   C. Both A and B
   D. Neither A nor B

5. All of the following are part of a truck brake system **EXCEPT:**
   A. drums.
   B. gladhands.
   C. reservoirs.
   D. rims.

6. All of the following are part of a truck's hydraulic braking system **EXCEPT:**
   A. the master cylinder.
   B. the governor.
   C. the wheel cylinders.
   D. the metering valve.

7. Which would be the LEAST likely cause of an antilock brake system (ABS) failure?
   A. A failed axle sensor
   B. A failed electronic control unit
   C. A defective modulator assembly
   D. A chipped wheel sensor tooth

8. Which of the following would be the LEAST likely cause of wheel lockup?
   A. Improperly adjusted brakes
   B. Leaking service chamber
   C. Bald tires
   D. Aggressive braking on slippery roads

9. The most likely cause of low system air pressure in a truck air brake system would be:
   A. an out-of-adjustment governor
   B. cracked supply reservoir
   C. leaking treadle valve
   D. a defective ratio valve

10. The most likely result of a complete secondary circuit failure in a truck air brake system would be:
    A. full wheel lockup
    B. complete brake failure
    C. visible and audible driver alert
    D. loss of trailer brakes

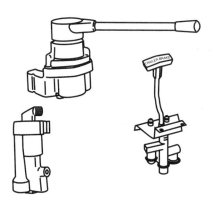

11. In the figure above, the function of the control valve is to:
    A. provide tractor front-wheel braking
    B. enable smooth bobtail stops
    C. apply trailer parking brakes
    D. apply trailer service brakes

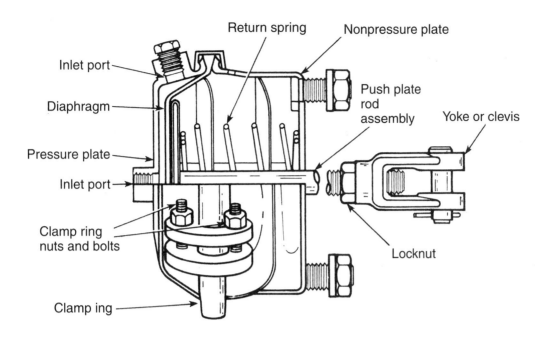

12. In the figure above, observe the air brake chamber. Technician A says that this assembly should never be disassembled without first caging the spring. Technician B says that this type of brake chamber is used only for parking brake applications. Who is right?
    A. A only
    B. B only
    C. Both A and B
    D. Neither A nor B

Answers to the Test Questions for the Pretest

1. B, 2. B, 3. A, 4. A, 5. D, 6. B, 7. D, 8. B, 9. A, 10. C, 11. D, 12. D

Explanations to the Answers for the Pretest

Question #1
Answer A is wrong.
Answer B is correct. The air compressor on a truck air brake system is driven by the vehicle engine either directly by a gear or by a belt and pulley.
Answer C is wrong.
Answer D is wrong.

Question #2
Answer A is wrong. The term reservoir can be used to describe any air tank in the system.
Answer B is correct. The term wet tank is commonly used to describe the supply reservoir.
Answer C is wrong.
Answer D is wrong. The primary reservoir is located downstream and is supplied by the supply reservoir.

Question #3
Answer A is correct. A defective spring brake assembly can fail to release and lock a wheel up.
Answer B is wrong. There is no parking brake cable on a truck air brake system.
Answer C is wrong. This answer is wrong because only Technician A is correct.
Answer D is wrong. This answer is wrong because Technician A is correct.

Question #4
Answer A is correct. Federal law requires all trucks over 10,000 GVW be equipped with a dual-circuit hydraulic brake system.
Answer B is wrong. Federal law sets specific brake system standards for all highway vehicles.
Answer C is wrong. This answer is wrong because only Technician A is correct.
Answer D is wrong. This answer is wrong because Technician A is correct.

Question #5
Answer A is wrong. Drums are part of a typical truck brake system.
Answer B is wrong. Gladhands are used to couple the trailer air brake supply and service air lines.
Answer C is wrong. Reservoirs are used to store the compressed air required for truck air brake systems.
Answer D is correct. Rims are considered to be part of the wheel assembly.

Question #6
Answer A is wrong. The master cylinder is part of a truck hydraulic brake system.
Answer B is correct. There is no governor in a hydraulic brake circuit.
Answer C is wrong. Wheel cylinders are part of a hydraulic brake circuit.
Answer D is wrong. Metering valves are part of a truck hydraulic brake system.

Question #7
Answer A is wrong. A failed axle sensor will cause an ABS malfunction.
Answer B is wrong. A failed electronic control unit will cause an ABS failure.
Answer C is wrong. A defective modulator assembly will cause an ABS failure.
Answer D is correct. A chipped wheel sensor tooth will not necessarily cause a complete ABS system failure and therefore this is the least likely option.

Question #8
Answer A is wrong. Improperly adjusted brakes can cause brake torque imbalance that will result in wheel lockups.
Answer B is correct. A leaking service chamber will result in brake performance problems but will not cause wheel lockup.
Answer C is wrong. Bald tires can contribute to a wheel lockup condition.
Answer D is wrong. Aggressive braking can cause wheel lockup on both dry and slippery pavement.

Question #9
Answer A is correct. System pressure is managed by the governor.
Answer B is wrong. A cracked supply reservoir will produce a complete system failure.
Answer C is wrong. A leaking treadle valve will cause system problems that would be reported before these affected system pressure.
Answer D is wrong. A ratio valve proportions front axle braking force.

Question #10
Answer A is wrong. A total secondary circuit failure alone cannot cause wheel lockup.
Answer B is wrong. In most cases when a secondary circuit failure occurs, the vehicle can be brought to a halt with no reduction in brake performance.
Answer C is correct. When a secondary circuit failure occurs, Federal Motor Vehicle Safety Standards (FMVSS) 121 requires that there be a visible driver warning that requires him to bring the vehicle safely to a standstill.
Answer D is wrong. System two-way check valves make it unlikely that brake performance anywhere on the tractor-trailer combination will be immediately affected in the event of a secondary circuit failure.

Question #11
Answer A is wrong.
Answer B is wrong.
Answer C is wrong. The trailer control valve should NEVER be used to park a rig.
Answer D is correct. The trailer control valve or spike isolates the trailer service brakes from the rest of the rig and is used for trailer-only service applications.

Question #12
Answer A is wrong. The figure shows a service brake chamber which contains only a light tension retraction spring: the diaphragms in service brake chambers can be safely serviced.
Answer B is wrong. This type of brake chamber is only used for service braking in current systems.
Answer C is wrong. The statements by both technicians are incorrect.
Answer D is correct. The statements by both technicians are incorrect, so this makes this answer correct.

Types of Questions

ASE certification tests are often thought of as being tricky. They may seem to be tricky if you do not completely understand what is being asked. The following examples will help you recognize certain types of ASE questions and avoid common errors.

Each test is made up of forty to eighty multiple-choice questions. Multiple-choice questions are an efficient way to test knowledge. To answer them correctly, you must think about each choice as a possibility, and then choose the one that best answers the question. To do this, read each word of the question carefully. Do not assume you know what the question is about until you have finished reading it.

Multiple-Choice Questions

One type of multiple-choice question has three wrong answers and one correct answer. The wrong answers, however, may be almost correct, so be careful not to jump at the first answer that seems to be correct. If all the answers seem to be correct, choose the answer that is the most correct. If you readily know the answer, this kind of question does not present a problem. If you are unsure of the answer, analyze the question and the answers. For example:

Question 1:

When considering a tractor service brake relay valve:
A. air lines are connected from the service brake relay valve delivery ports to the rear axle service brake chambers.
B. when the brakes are leased the inlet valve is open in the service brake relay valve.
C. when the brakes are released the exhaust valve is closed in the service brake relay valve.
D. the service brake relay valve is in a balanced position when the air pressure in the rear brake chambers equals reservoir pressure.

Analysis:

Answer A is correct. Air lines are connected from the service brake relay valve delivery ports to the rear axle service brake chambers.
Answer B is wrong. When the brakes are released the inlet valve is closed, not open, in the service brake relay valve.
Answer C is wrong. When the brakes are released the exhaust valve is open—not closed—in the service brake relay valve.
Answer D is wrong. The service brake relay valve is in a balanced position when the air pressure in the rear brake chambers equals system pressure.

EXCEPT Questions

Another type of question used on ASE tests has answers that are all correct except one. The correct answer for this type of question is the answer that is wrong. The word "EXCEPT" will always be in capital letters. You must identify which of the choices is the wrong answer. If you read quickly through the question, you may overlook what the question is asking and answer the question with the first correct statement. This will make your answer wrong. An example of this type of question and the analysis is as follows:

Question 2:
> When diagnosing a service brake relay valve, all of the following apply **EXCEPT:**
> A. inlet valve leakage is tested with the service brakes released.
> B. apply a soap solution to the area around the inlet and exhaust valve retaining ring to check exhaust valve leakage.
> C. exhaust valve leakage is tested with the brakes applied.
> D. the control port on the service brake relay valve is connected to the supply port on the brake application valve.

Analysis:

Answer A is wrong. The inlet valve leakage is tested with the service brakes released.
Answer B is wrong. You do apply a soap solution to the area around the inlet and exhaust valve retaining ring to check exhaust valve leakage.
Answer C is wrong. Exhaust valve leakage is tested with the brakes applied.
Answer D is correct. This is the exception because the control port on the service brake relay valve is connected to the delivery port on the brake application valve.

Technician A, Technician B Questions

The type of question that is most popularly associated with an ASE test is the "Technician A says... Technician B says. . . . Who is right?" type. In this type of question, you must identify the correct statement or statements. To answer this type of question correctly, you must carefully read each technician's statement and judge it on its own merit to determine if the statement is true.

Typically, this type of question begins with a statement about some analysis or repair procedure. This is followed by two statements about the cause of the problem, proper inspection, identification, or repair choices. You are asked whether the first statement, the second statement, both statements, or neither statement is correct. Analyzing this type of question is a little easier than the other types because there are only two ideas to consider although there are still four choices for an answer.

Technician A. . . Technician B questions are really double-true-false questions. The best way to analyze this kind of question is to consider each technician's statement separately. Ask yourself, is A true or false? Is B true or false? Then select your answer from the four choices. An important point to remember is that an ASE Technician A... Technician B question will never have Technician A and B directly disagreeing with each other. That is why you must evaluate each statement independently. An example of this type of question and the analysis of it follows.

Question 3:
> While discussing single and double check valves, Technician A says single check valves are connected between the primary and secondary reservoirs. Technician B says a double check valve allows air pressure to flow from the lowest of two pressure sources. Who is correct?
> A. A only
> B. B only
> C. Both A and B
> D. Neither A nor B

Analysis:

Answer A is correct. Single check valves are connected between the primary and secondary reservoirs.
Answer B is wrong. A double check valve allows air pressure to flow from two different locations.
Answer C is wrong.
Answer D is wrong.

Questions with a Figure

About 10 percent of ASE questions will have a figure, as shown in the following example:

Question 4:

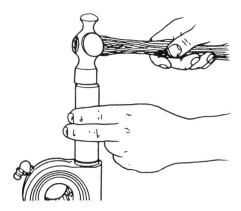

As shown in the figure above, you are installing a lip seal on a worm gear of a slack adjuster. Technician A says to install the lip of the seal towards the outside of the adjuster. Technician B says not to hit the seal after it reaches the bottom of the bore. Who is right?

A. A only
B. B only
C. Both A and B
D. Neither A nor B

Analysis:

Answer A is wrong. Answer A is a good choice because it says you install the lip of the seal towards the outside of the adjuster. Yet, it is wrong because both technicians are right.

Answer B is wrong. Answer B is a good choice because you do not have to hit the seal after it reaches the bottom of the bore. Yet, it is wrong because both technicians are right.

Answer C is correct.

Answer D is wrong.

Most-Likely Questions

Most-likely questions are somewhat difficult because only one choice is correct while the other three choices are nearly correct. An example of a most-likely-cause question is as follows:

Question 5:

During a brake inspection, a technician tests the air supply system and finds that the buildup is slow. Which of these is the most likely cause?

A. A clogged compressor inlet filter
B. A leak in a brake chamber
C. An air leak in the cab suspension
D. A restriction in the governor line

Analysis:
Answer A is correct. A clogged compressor inlet filter is the most likely cause of slow buildup.
Answer B is wrong. A leak in a brake chamber is not as likely a cause for slow build-up.
Answer C is wrong. An air leak in the cab suspension is an unlikely cause.
Answer D is wrong. A restriction in the governor line will cause high pressure.

LEAST-Likely Questions

Notice that in most-likely questions there is no capitalization. This is not so with least-likely type questions. For this type of question, look for the choice that would be the least likely cause of the described situation. Read the entire question carefully before choosing your answer. An example is as follows:
Question 6:
Which of these is the LEAST likely cause of wheel bearing failure?
A. overloading
B. contamination
C. a damaged race/axle housing
D. improper lubricant

Analysis:
Answer A is correct. Overloading is the least likely cause of high heat buildup.
Answer B is wrong. Contamination will cause high heat buildup, destroying the bearing.
Answer C is wrong. Axle housing damage may cause wheel bearings to fail.
Answer D is wrong. Improper or incompatible lubricant will cause failure of the bearing.

Summary

There are no four-part multiple-choice ASE questions having "none of the above" or "all of the above" choices. ASE does not use other types of questions, such as fill-in-the-blank, completion, true-false, word-matching, or essay. ASE does not require you to draw diagrams or sketches. If a formula or chart is required to answer a question, it is provided for you. There are no ASE questions that require you to use a pocket calculator.

Testing Time Length

An ASE test session is four hours and fifteen minutes. You may attempt from one to a maximum of four tests in one session. It is recommended, however, that no more than a total of 225 questions be attempted at any test session. This will allow for just over one minute for each question.

Visitors are not permitted at any time. If you wish to leave the test room, for any reason, you must first ask permission. If you finish your test early and wish to leave, you are permitted to do so only during specified dismissal periods.

Monitor Your Progress

You should monitor your progress and set an arbitrary limit to how much time you will need for each question. This should be based on the number of questions you are attempting. It is suggested that you wear a watch because some facilities may not have a clock visible to all areas of the room.

Registration

Test centers are assigned on a first-come, first-served basis. To register for an ASE certification test, you should enroll at least six weeks before the scheduled test date. This should provide sufficient time to assure you a spot in the test center. It should also give you enough time for study in preparation for the test. Test sessions are offered by ASE twice each year, in May and November, at over six hundred sites across the United States. Some tests that relate to emission testing also are given in August in several states.

To register, contact Automotive Service Excellence/American College Testing at:

ASE/ACT
P.O. Box 4007
Iowa City, IA 52243

4 An Overview of the System

Brakes (Test T4)

This section includes the task areas and task lists for this test and a written overview of the topics covered in the test.

The task list describes the actual work you should be able to do as a technician and that you will be tested on by the ASE. This is your key to the test and you should review this section carefully. We have based our sample test and additional questions upon these tasks, and the overview section will also support your understanding of the task list. ASE advises that the questions on the test may not equal the number of tasks listed; the task lists tell you what ASE expects you to know how to do and to be ready to be tested on.

At the end of each question in the Sample Test and Additional Test Questions section, a letter and number will be used as a reference back to this section for additional study. Note the following example: **A1.1.**

Task List

A. Air Brakes Diagnosis and Repair (38 Questions)

1. Air Supply and Service Systems (17 Questions)

Task A1.1 Inspect, repair, or replace system pressure controls (governor/relief valve), unloader assembly, valves, filters, line, hoses, and fittings.

Example:
1. Technician A states that the difference between governor cut-in and governor cut-out must not exceed 25 psi. Technician B states that when the difference between governor cut-in and governor cut-out exceeds 25 psi, the governor must be adjusted. Who is right?
   A. A only
   B. B only
   C. Both A and B
   D. Neither A nor B (A1.1)

Question #1
Answer A is correct because Federal Motor Vehicle Safety Standards (FMVSS) 121 requires that the difference between governor cut-in and governor cut-out must not exceed 25 psi.
Answer B is wrong because the only adjustment on an air system governor is the governor cut-out value. When the difference between governor cut-in and governor cut-out exceeds 25 psi, the governor is defective and must be replaced.
Answer C is wrong because only one technician is correct.
Answer D is wrong because one technician is right.

Task List and Overview

A. Air Brakes Diagnosis and Repair (38 Questions)

1. Air Supply and Service Systems (17 Questions)

Task A1.1 Diagnose poor stopping, premature wear, brake noise, air leaks, pulling, grabbing, or dragging problems caused by supply and service system malfunctions; determine needed repairs.

The potential energy of an air brake system is compressed air: the action of the driver's foot on a brake pedal contributes nothing to developing this potential energy which is developed by an engine-driven air compressor.

A tractor-trailer air brake system can be divided into three circuits: the supply circuit, the control circuit, and the foundation assembly. The supply circuit is responsible for making compressed air at the correct system pressure available to the control circuit. The control circuit manages the service and parking brake functions of the vehicle. The foundation brakes are the mechanical shoes, drums, rotors, and pads that stop the truck and are managed by the control circuit. A failure within any of these three circuits can cause anything from a minor malfunction to a complete brake failure.

Truck brakes are required to be balanced. A balanced braking system can be defined as one in which the braking pressure reaches each actuator at the same moment and at the same pressure level. Factors that affect brake balance are application and release times. To meet balanced performance requirements, vehicle manufacturers match all the brake system valves and components including the hose size and fitting geometry used in the system. Air application and release performance is dependent on both the size and volume of chambers and the distance the air must travel.

In inspection for leaks in an air system is an important part of any brake system inspection. Every component and the entire plumbing circuit of the air brake system should be checked.

Air-related causes of dragging brakes on a truck air brake system can include a leaking hold-off diaphragm, low holdoff pressure in the spring brake section of the brake chamber, spring brake control valve problems, low system pressure causing partial application of the spring brakes, and sticking service application and relay valves.

The majority of braking on a typical highway involves application pressures of 20 psi or less. Full application pressure stops are rare must be capable of them. A truck air system must be able to accommodate 4–6 full reserve stops with the brakes properly adjusted.

Brakes that are mechanically out of adjustment require a greater volume of application air.

Unlike hydraulic fluid, air is compressible. The larger the volume of air, the higher its compressibility. So brake timing lags are greater in air brake systems than in hydraulic brake systems. When a brake control signal has to travel from the application valve in the tractor to an actuator valve in the trailer, the lag time depends on the distance the signal has to travel. Tractor-trailer brake systems compensate for this by pneumatically balancing the relay valves in the system by crack pressure: the valves furthest from the application valve are designed with the lowest crack pressures, with those located on the tractor itself having slightly higher crack pressure values. The objective is to have simultaneous application of the brakes in each wheel assembly on the rig at both low- and high-application pressures. Current FMVSS 121 brake application timing requirements:

    Trailer: Application: 0.39 seconds   Release: 0.65 seconds
    Truck: Application: 0.45 seconds   Release: 0.55 seconds

Pneumatic imbalance may cause uneven braking in a truck brake system. Causes of pneumatic imbalance may be defective valves or plumbing irregularities such as restricted lines, the fitting of a 90° fitting in place of a 45° fitting, or using brake hose with smaller- or larger-than-specification inside diameters.

To perform a full pressure balance and pressure buildup (timing) test, a technician will need a pair of test hoses and a duplex gauge. For multiple combinations, two additional hoses will be needed per trailer. Ideal balance timing in a combination vehicle is defined as each axle receiving identical air pressure simultaneously on a brake application.

Task A1.2   Check air system buildup time; determine needed repairs.

System air buildup times are defined by federal legislation, specifically FMVSS 121. This legislation defines the required buildup times and values. A common check performed by enforcement agencies requires that the supply circuit on a vehicle be capable of raising air system pressure from 85–100 psi in 25 seconds or less.

Failure to achieve this buildup time indicates a worn compressor, defective compressor unloader assembly, supply circuit leakage, or a defective governor.

Task A1.3   Drain air reservoir tanks; check for oil, water, and foreign material; determine needed repairs.

Air reservoirs store and provide air for the truck air braking system. Servicing of the reservoirs consists of inspection, draining the tanks, and performing leakage tests. Air brake systems must have at least three air reservoirs. They may, and usually do, have more. Compressed air from the compressor is delivered to the supply reservoir. The supply reservoir is also known as a wet tank. The supply reservoir feeds air to primary and secondary reservoirs which supply the brake circuit with air.

A wet tank or supply tank is so named for the moisture that forms in such a tank when hot compressed moisture-laden air cools and condenses on the tank inside walls.

Reservoirs may be equipped with either automatic or manual drain valves. Daily draining of manual drain valves is recommended to keep the system free of contaminants and moisture. Automatic reservoir drain valves should be checked for proper operation, also on a daily basis.

Supply reservoirs are equipped with a safety valve also known as a pop-off valve. This valve is designed to trip at 150 psi. This protects the system in the event of a governor failure.

Air reservoirs on trucks are pressure vessels. They are hydrostatically tested after manufacture. The inside wall is treated with a corrosion-protection coating. Truck air reservoirs should never be repair-welded, first because they are pressure vessels and second because the internal coating is destroyed. Leaking tanks should be replaced.

Evidence of oil in a wet tank often indicates the compressor is pumping oil through the system. Oil can damage valving throughout the system, so the source of oil contamination must be determined immediately and repaired.

Task A1.4 Inspect, adjust, align, or replace compressor drive belts and pulleys.

Some compressors, especially those on smaller engines or in those cases where an air brake system has been retrofit, have air compressors driven by the engine using a belt and pulley. Belt sets, pulleys, and idlers should be routinely inspected for indications of wear and axial runout. Belt tension should be set to specification using a belt tension gauge. Cracks and nicks in a drive belt require that it be replaced.

Task A1.5 Inspect, time, or replace compressor drive gear and coupling.

Most compressors are two-cylinder pumps that are balanced units: these do not have to be timed to the engine on installation. Ensure that the oil feed tube is correctly aligned in the compressor crankshaft and that the gear teeth are not damaged.

Some compressors are not self-balanced units, usually single-cylinder types, and these must be timed to the engine on installation. Ensure that the manufacturer's service literature is consulted when installing compressors that must be timed to the engine.

Task A1.6 Inspect, repair, or replace air compressor, air cleaner, oil supply, water lines, hoses, and fittings.

A compressor is an air pump. The basic air compressor operates much like an internal combustion engine. It consists of a crankcase, cylinder block, and cylinder head. The crankshaft is supported in the crankcase by main bearings. Connecting rods are connected to the throws on the crankshaft at their big end. The piston assemblies are connected to the small end of the crankshaft by means of a wrist pin. As the crankshaft is rotated, the pistons reciprocate in the cylinder bores in the compressor block. Piston rings seal the piston in the cylinder bore and control an oil film on the cylinder wall.

The cylinder head assembly is equipped with discharge valves and discharge ports. It also has reed-type inlet valves. On the piston downstroke, the inlet valve is unseated by lower pressure in the cylinder, and a charge of air is induced into the cylinder. When the piston passes through the bottom of its travel to be driven upward, the inlet valves are seated by the cylinder pressure. As the piston continues to be driven upward, pressure in the cylinder rises, and when it exceeds the pressure in the discharge port, the discharge valve opens and compressed air is forced out of the cylinder into the discharge line.

An unloading mechanism in the cylinder head is controlled by a governor. The governor uses an air signal to cycle the compressor through effective (pumping) and unloaded (not pumping) cycles. The air signal from the governor acts on the unloader assembly which holds the inlet valves open throughout the cycle. When this occurs, the compressor is unable to compress air as the cylinder cannot be sealed; it is therefore unloaded.

The compressor is pressure-lubricated by the engine lubrication system. Pressurized oil is usually delivered to the compressor crankshaft by means of a tube from a bore in the drive gear. The crankshaft main bearings, crankshaft, and piston assemblies require lubrication. Compressors are liquid-cooled by the engine cooling system through inlet and outlet lines located in the cylinder head. The act of compressing air heats it up. Compressors therefore are required to be cooled to keep air discharge temperatures at 300° F or less.

When troubleshooting a noisy compressor, a technician must determine the source of the noise. Squeaking drive belts or pulleys, failed bearings, and lubrication-related problems are often the cause of a noisy compressor.

An air compressor is the source of compressed air to all the vehicle pneumatic components, including windshield wipers, steering assist units, suspensions, and air starters.

Task A1.7 Inspect, test, adjust, or replace system pressure controls (governor/relief valve), unloader assembly valves, filters, lines, hoses, and fittings.

Many different air valves are used to control, regulate, or condition the air in a braking system. These valves guide the direction of flow, control the amount of pressure, and remove contaminants from the air system(s) in which they serve.

The governor manages system pressure. It monitors pressure in the supply tank by means of a line directly to it. This pressure acts on a diaphragm and spring within the governor. The spring tension is adjustable. The governor manages compressor-loaded and unloaded cycles. Loaded cycle is the compressor-effective cycle—that is, the compressor is pumping air. Unloaded cycle is when the compressor is being driven by the engine but not actually compressing air; it induces a charge of air through the inlet ports on piston downstroke and pushes it out through the same inlet port through the upstroke.

When a compressor is driven by the engine, it is in its loaded cycle until it receives an air signal from the governor to put it into unloaded cycle. This signal acts on the unloader assembly in the compressor cylinder head. The function of the unloader assembly is to hold the inlet valves open.

The governor simply controls whether the compressor is in loaded or in unloaded operating mode. It defines the system pressure. System pressure in most trucks is set at values between 110 and 130 psi, with 120 psi being typical. System pressure is known as cut-out pressure, the pressure at which the governor outputs the unloader signal to the compressor. The unloader signal is maintained until pressure in the supply tank drops to the cut-in value. Cut-in pressure is required by FMVSS 121 to be no more than 25 psi less than that of the cut-out value. The difference on most systems ranges between 20 and 25 psi.

Governor operation can be easily checked. One method is to drop the air pressure in the supply tank to below 60 psi and with the vehicle's engine running, build the pressure. A master gauge should be used to record the cut-out pressure value. This should be exactly at the specification value. If not, remove the dust boot at the top of the governor, release the locknut, and turn the adjusting screw either clockwise or counterclockwise, to lower or raise the cut-out pressure.

Governors seldom fail but they do not have check valves. If the unloader signal is not delivered to the compressor unloader assembly, high system pressures will result. If the safety pop-off valve on the supply tank trips (this occurs at 150 psi), this is usually an indication of governor or compressor unloader malfunction.

The only adjustment on an air governor is the cut-out pressure value. If the difference between governor cut-out and cut-in is out of specification, the governor must be replaced.

Task A1.8 Inspect, repair, or replace air system lines, hoses, fittings, and couplings.

Brake hose must be replaced to the original standard, or brake performance might be compromised.

Sizing of lines on both the tractor and the trailer will affect both application and release timing of the brakes. Every fitting used in brake system plumbing affects the fluid dynamics and care should be taken to always replace fittings with those similar to the original ones. Replacing a 45° elbow with a 90° elbow is equivalent to adding 7 additional feet of brake hose, and this will affect pneumatic timing.

Brake hose should be securely clamped away from moving components when installed. When reusing dryseal fittings, the nipples and seats should be inspected. Department of Transportation (DOT) approved hose should be used when replacing defective hose.

Brake hose can fail internally, sometimes forming a rubber flap that can act as a check in the line, permitting air to flow towards a valve but trapping it there.

Task A1.9 Inspect, test, clean, or replace air tank relief (pop-off) valves, check valves, drain cocks, spitter valves, heaters, wiring, and connectors.

A system safety valve is located usually in the supply tank. It is designed to trip at a pressure value of 150 psi. It is nonadjustable and consists of a ball seat and spring. Its function is to relieve system air if the pressure builds to a dangerously high level, such as would occur when a governor failed.

The supply tank feeds the primary and secondary reservoirs of the brake system. Each is pressure-protected by means of a one-way check valve. One-way check valve operation can be verified by removing the supply and checking back leakage.

One-way check valves are used variously throughout an air brake system to pressure-protect and isolate portions of the circuit.

Automatic drain plugs can become plugged with sludge (oil and water residues) and may require periodic cleaning. Should oil and excessive water be evident at the drain plugs, check the air dryer and/or compressor.

Task A1.10 Inspect, test, clean, repair, and replace air drier systems, filters, valves, heaters, wiring, and connectors.

Moisture in an air system can be very damaging. Ambient moisture is a problem any time the relative humidity is high, so it is both a summer and a winter problem. The airborne moisture condenses in the reservoirs as the compressed air cools. Most current systems use air dryers to remove moisture from the compressed air before it gets to the supply tank. These use two principles to remove moisture from the compressed air. The first type is the desiccant type. During the charge cycle, hot compressed air passes through a desiccant pack that adheres the moisture; dry air exits to the discharge port and the supply reservoir. At governor cut-out, an air signal from the governor begins the purge cycle of the air dryer. This acts on the purge piston, which exhausts any water collected in the purge orifice to atmosphere. The second type is the heat exchanger type. The heat exchanger type of air dryer attempts to cool the compressed air to the point that the moisture is condensed; once condensed, it can be separated and dumped by means of a purge valve. Some air dryers use a combination of both principles.

Many air dryers use a heater to prevent icing in cold weather; the heater is thermostat regulated. In cold weather, the thermostat controls the heater cycles to maintain a temperature exceeding 45° F.

Oil can destroy the desiccant pack in an air dryer. When an air compressor fails and pumps its lubricating oil through the system, the desiccant pack becomes contaminated and requires replacement.

Task A1.11 Inspect, test, adjust, repair, or replace brake application (foot) valve, fittings, and mounts.

The service application valve in a truck brake system is a floor-mounted foot valve, known as a treadle valve. The treadle valve is actually two valves in one. The upper portion of the valve is the primary section and the lower portion is the secondary or relay section. Each section has a dedicated feed and its own exhaust port. The upper or primary section of the treadle valve is supplied directly from the primary reservoir. The lower or relay section of the valve is supplied by the secondary reservoir.

The treadle valve is actuated mechanically, that is, by foot pressure from the driver. When the driver's foot acts on the treadle valve, the primary piston is forced downward. This movement first closes the primary exhaust port and then modulates air proportional to piston travel to:

1. actuate whatever brakes are plumbed into the primary circuit; this is usually, but not exclusively, the drive axles on a typical tandem drive tractor unit.
2. actuate the relay or secondary piston, located below the primary section in a dual-circuit application valve.
3. act against the mechanical pressure (foot pressure) to provide brake feel.

The secondary or relay section of the dual-circuit application valve is actuated pneumatically by primary circuit air. This section operates similarly to a relay valve in that a signal pressure value (the air from the primary section) is used to displace a relay piston that then modulates secondary circuit air to whatever brakes/valves are located in the secondary circuit. In a typical tractor air brake system, this would normally be the front axle brakes. Like a relay valve, secondary section is designed to modulate an air pressure value to the secondary circuit, identical to the signal pressure.

Each portion of the valve has its own exhaust port. Primary and secondary circuit air never come into contact with each other in the treadle valve.

In an emergency application of the treadle valve, both the primary and secondary inlet valves are held open and full reservoir pressure is applied to each of the two circuits.

In the event of a total primary circuit failure, no air is available to the primary section of the treadle valve. The driver low pressure warning alerts would trip and when the treadle valve is depressed, foot pressure would drive the primary piston downward until it mechanically contacts the relay piston to actuate the secondary circuit. There would be zero brake "feel" and there would be greater pedal travel.

In the event of a total secondary circuit failure, the primary section of the treadle valve would function normally. However, the driver low air pressure alert would trip and the vehicle would have to be brought to a halt using primary circuit source air only.

Task A1.12 Inspect, test, or replace two-way (double) check valves and anti-compounding valves.

Two-way (double) check valves play an important role as a safeguard in dual circuit air brake systems. The typical two-way check valve is a T with two inlets and a single outlet. It outputs the larger of the two source pressures to the outlet port and checks the lower value source: the valve will shuttle in the event of a change in the source pressure value. In other words it will always prioritize the higher source pressure. Two-way check valves provide a means of providing the primary circuit with secondary circuit air and vice versa in the event of a circuit failure. In the event that both source pressures are equal, as would be the case in a properly functioning dual air brake circuit, the valve will prioritize the first source to act on it.

Compounding occurs when a brake foundation is subjected to both mechanical and pneumatic force. A spring brake chamber in park mode has the air acting on the holdoff chamber exhausted from it; this enables the mechanical force of the spring in the chamber to act on the slack adjuster to apply the brakes. As it takes an air pressure of approximately 60 psi to cage the spring brake, it is capable of applying approximately the equivalent of that amount of force to the foundation brakes when no air is acting on the holdoff diaphragm. If when in this parked condition a driver made a full service application of the brakes, the mechanical force that the spring was applying to the foundation brakes would be compounded by a further 120 psi acting on the service diaphragm. The result would be a total application pressure 50 percent greater than the specified maximum. To prevent this from happening, anti-compounding valves are used. These operate by dividing service application air between the service and holdoff chambers whenever a service application is made in the absence of air in the holdoff chamber.

Anti-compounding valve operation can be easily verified with a pair of air pressure gauges fitted to the service and holdoff lines while the vehicle is parked.

Failure of an anti-compounding valve can result in twisted S-cam shafts, spline damage, and slack adjuster damage.

Task A1.13 Inspect, test, repair, or replace stop and parking brake light circuit switches, wiring, and connectors.

Functioning brake lights are required in all highway vehicles. Trucks with air brakes require two brake light switches, either of which can illuminate the vehicle brake light circuit.

The service stop light switch is a normally open, air-actuated switch, plumbed into the service brake circuit. The switch is closed by a small application pressure acting on it.

The parking brake light switch works oppositely. This switch is designed to close the electrical circuit when no air is acting on it and to open when air is charged to the holdoff chambers.

The operation of both switches can be verified with a digital multimeter (DMM). The voltmeter should be used when testing either switch in circuit and the ohmmeter, when out of circuit.

Task A1.14 Inspect, test, repair, or replace hand brake (trailer) control valve, lines, hoses, fittings, and mountings.

The trailer control valve is used to actuate the trailer service brakes independently of the tractor's service brakes. This valve is also known as trolley valve, broker brake, and spike. The source of air supplied to the trailer control valve is usually the secondary circuit. This system pressure air is modulated proportionally with control valve travel.

Two-way check valves are used in the application circuit of the trailer brakes. The two-way check valve is located downstream from both the trailer control valve and the treadle valve. In a typical system, the trailer application brake valve (trolley or spike) uses secondary circuit sourced air and the treadle valve trailer service signal uses primary circuit sourced air to activate the trailer service brakes. If a driver was using the spike to bring the vehicle to a stop and during this process an emergency occurred that required a panic application of the treadle valve, the two-way check valve would permit the source air to change from secondary circuit air from the trailer control valve to the higher source pressure value from the treadle valve.

Trailer control valves performance can be verified with an accurate air pressure gauge. The trailer is supplied with air by means of air hoses. Service and trailer supply air is plumbed to a tractor protection valve. This air is transferred to the trailer by couplers known as gladhands. Gladhands enable easy coupling between the tractor air supply and controls and the trailer pneumatic circuit.

Gladhand seals are usually manufactured out of rubber. They are retained in an annular groove in the gladhand and are easily replaced when they fail.

Task A1.15 Inspect, test, or replace brake relay valve.

Relay valves permit a remote air signal (from the treadle or trailer valve) to effect service braking with an air supply close to the brake chambers. The valve is controlled by a signal that is plumbed to its service port. At the supply port, system air pressure is available to the valve from a close-by air reservoir. When the relay valve receives an air signal, this pressure acts on the relay piston. When the relay piston is displaced, it modulates the air available at its supply port, to actuate the service brake chambers connected to its delivery ports. One relay valve can typically manage two to four service brake chambers. The valve is usually designed so that the signal value is identical to the output value modulated to the delivery ports. When the signal pressure is dropped or relieved, the relay valve exhausts the service supply air returning from the service chambers, and a retraction spring returns the relay valve to its neutral position.

A sticking relay piston in the relay valve will cause supply air to be modulated to the service chambers when there is no signal pressure; this causes dragging brakes or locked brakes depending on the location of the relay piston. Moisture in the air system may freeze and prevent the relay retraction spring from returning the relay piston.

Relay valve operation can be verified using a pair of gauges that monitor signal pressure and delivery.

Task A1.16 Inspect, test, and replace quick-release valves.

Quick-release valves are mounted close to the brake chambers or components they serve. They are used throughout the air brake system and other vehicle air systems to speed release times.

The typical quick-release valve has a single inlet port and two outlet ports. When air is charged to the inlet port, the valve acts like a T fitting and simply divides the air to the two outlets. However, when the air pressure supplied to the inlet is exhausted at the application or control valve, the air being outputted from the quick release to the brake chambers is exhausted at the quick-release valve.

A quick-release valve may also be used on the service gladhand supplying the trailer(s). This feature greatly reduces the release lag that can occur in service braking, by exhausting the service signal at a much higher speed.

Because of its simplicity, quick-release valves seldom fail. They are vulnerable to external damage because of their location and contaminated air, which may cause exhaust port leakage.

Task A1.17 Inspect, test, and replace front and rear axle limiting (proportioning) valves.

All trucks today are required to have front-wheel brakes. Some much older trucks had a manual front axle limiting valve that enabled the driver to toggle between dry roads for full front-wheel braking or slippery roads to reduce by 50 percent application pressure to the front-wheel brakes. Today's trucks are fitted with automatic front-wheel proportioning valves also known as ratio valves. A ratio valve manages air pressure to the front brake chambers by proportioning the application pressure delivered to it. Typically, the ratio valve will proportion as follows:

Application pressure	Pressure to front brakes
0–10 psi	0
10–40 psi	50% application pressure
40–60 psi	50% graduating to 100%
60 psi and higher	100% application pressure

The bobtail proportioning valve senses when the tractor is bobtailing and operates when the tractor parking brake is released and the trailer supply valve remains closed. The objective is to reduce the air pressure applied to the tractor's drive axle(s) and minimize drive wheel lockup on slippery pavement. The Bobtail proportioning valve is integral with the relay valve in managing the service braking. This valve will function as a relay valve under normal operation. In bobtail mode, the valve proportions the air to the brake chambers by bleeding a percentage of the application air to exhaust.

Task A1.18 Inspect, test, and replace tractor protection valve.

The tractor protection valve protects the tractor air supply under a trailer breakaway condition or severe air leakage. Two air lines connect a tractor brake system with a trailer air circuit. The trailer supply line supplies the trailer with air for braking and for any other pneumatic systems such as its suspension. The trailer service line is the service brake signal line. Both these lines are plumbed to the tractor protection valve. When the trailer supply dash valve (emergency) is in its off position, the tractor protection valve is also off; in this condition, when the driver makes a service brake application, no air will exit the tractor protection valve service signal line to the trailer. When the trailer air supply valve is open, this air will open the tractor protection valve and supply the trailer pneumatic systems with air; in this condition, when the driver makes a service brake application, the service brake signal air will be transmitted through the tractor protection valve to apply its service brakes.

It should be noted that the trailer air supply controls the trailer park brakes. Whenever this air supply is interrupted, whether intentionally such as when the trailer is being parked or unintentionally, such as in a breakaway, the spring brakes will apply regardless of how much pressure is in the trailer air reservoirs.

The tractor protection valve is designed to isolate the tractor air system from that of the trailer at a predetermined value that ranges from 20 to 45 psi, depending on the system. It should be noted that at these pressures, the spring brakes on both the tractor and trailer would be partially applied.

Task A1.19 Inspect, test, and replace emergency (spring) brake control valve(s).

The spring brake valve supplies air to the holdoff chambers to release the spring brakes and enable the vehicle to move. During normal operation, it limits the holdoff pressure to the spring brake chambers to a value of around 90 psi; this speeds application times in the event of an emergency and permits a consistent holdoff pressure (system pressure fluctuates between cut-in and cut-out pressure values).

Some systems use an inversion valve. In the event of a primary air system failure, the inversion valve allows for a modulated application of the emergency air brake chambers by relieving air from the holdoff chamber in direct proportion to the application pressure at the treadle valve.

Spring brake control valves are often incorporated in multifunction valves that contain parking, service, and inversion functions.

Task A1.20 Inspect, test, or replace low-pressure warning devices.

FMVSS 121 requires that a driver receive a visible alert when the system pressure drops below 60 psi; in most cases this is accompanied by an audible alert, usually a buzzer. A low air pressure warning device is fitted to both the primary and secondary circuits. This is a simple electrical switch that can be plumbed anywhere into a system requiring monitoring. The switch is electrically closed whenever the air pressure being monitored is below 60 psi; when the air pressure value exceeds 60 psi, the switch opens.

Verifying the operation of a low air pressure warning switch can be done by pumping the service application valve until the system pressure drops to the trigger value.

Both the primary and secondary circuit air pressure must be monitored by a dash-located gauge. The required visible warning is usually a dash warning light.

Task A1.21 Inspect, test, repair, and replace air pressure gauges, lines, hoses, and fittings.

Air pressure gauge operation can be verified by using a master gauge, a good quality, liquid-filled gauge that uses a Bourdon principle of operation. When troubleshooting vehicle air pressure management problems, the vehicle gauges should not be relied on.

2. Mechanical/Foundation (11 Questions)

Task A2.1 Diagnose poor stopping, premature wear, brake noise, pulling, grabbing, or dragging complaints caused by foundation brake components, slack adjuster, and brake chamber problems; determine needed repairs.

The S-cam, wedge, or disc brake mechanism itself, linings or pads, and related parts such as brake chamber(s), slack adjuster(s), and parking brake components generally make up what is called the foundation brake assembly.

All vehicle braking requires that kinetic energy (the energy of motion) be converted to heat energy by means of friction; the heat energy must then be dissipated to atmosphere. The foundation brake assembly consists of those brake components that are responsible for effecting the retarding effort required to stop a vehicle.

In a typical S-cam, shoe/drum assembly, slack adjusters connect the brake actuation chambers with the foundation brake assembly. Slack adjusters are levers so their length is critical. The greater the distance between the centerline of the S-cam and the point at which the brake chamber clevis connects with the slack adjuster, the greater its leverage. Slack adjusters also convert the linear force produced by the brake chamber into torque (twisting force). Slack adjusters are simple components whose parts include a worm gear, a pushrod, an actuator, and an actuator piston. Current slack adjusters are required to be automatically adjusting.

The geometry of the relationship between the brake chamber and slack adjuster requires that the angle between the slack adjuster and the chamber pushrod be 90° when the brake is in the fully applied position. In any position other than 90°, the slack adjuster has less mechanical advantage. As the brake shoe friction surfaces wear, this 90-degree angle must be maintained by periodic adjustment of manual slack adjusters or functioning automatic slack adjusters.

Both parking and service brake performance relate directly to brake adjustment.

Foundation brake problems can cause unbalanced braking, grabbing, not releasing, or failure to apply.

Some causes of foundation brake problems are: use of improper replacement parts, cargo weight overloads, lining contamination, poorly maintained or installed brakes, broken or malfunctioning brake components, brakes out of adjustment, and overheating.

Task A2.2 Inspect, test, adjust, repair, or replace service brake chambers, diaphragm, clamp, spring, pushrod, clevis, and mounting brackets.

To service a spring brake chamber, it will be necessary to "cage" or release the compression spring before any service may be performed. A spring brake is normally manually released by tightening a nut on a supplied release tool. It is by this manner that air can then be released from the spring brake chamber, thus facilitating service on the spring brake.

If the clevis is not positioned correctly onto its pushrod, the brake will not completely release. A clevis has internally tapped threads for a pushrod to engage its threaded male end into. Most manufacturers furnish slack adjuster installation templates with units, thus facilitating correct installation.

A 0.060-inch gap between the clevis pin and the clevis collar is considered the maximum allowable before replacement. You replace the "clevis" if the collar hole is worn beyond this limit. The automatic slack adjuster automatically adjusts the clearance between the brake linings and the brake drum or rotor. The slack adjuster controls the clearance by sensing the length of the stroke of the pushrod for the air brake chamber, so no manual adjustment is required.

The more common causes of actively sluggish service braking in an air system are problems concerning foundation brake geometry. Problems such as an obstruction in a brake chamber or poor alignment of the brake linkage can and often times does produce apply- and release-rate performance problems.

To facilitate removing the spring chamber off the adapter, it is necessary to slide it sideways while holding the adapter. The diaphragm can be removed from the spring chamber by rotating it counterclockwise from the chamber. The clamps or any other part of the adapter or chambers should not be struck with a hammer or heavy object. Finally, the air pressure should be exhausted if it was used to aid in caging the spring brake.

Rotochambers were designed to replace the conventional type brake chamber. They operate similarly, and are identified by types. A type 30 Rotochamber would have an effective diaphragm area of 30 inches.

Spring brake chamber assemblies should not be repaired but replaced as an assembly, after they have been disarmed or manually released (caged) and discarded.

When the park control valve closes, the air in the rear portion of the spring chamber is exhausted, thus allowing the compression spring to apply the parking brake.

The pressure plate on a spring brake chamber is located between the adapter return spring and the compression spring. The pressure plate is what holds off the compression spring when the spring brake chamber is charged, thus releasing the spring (parking) brake. When air is released from the air chambers, the pushrod return spring, in combination with the brake shoe return spring, returns the diaphragm, push plate and pushrod assembly, slack adjuster, and brake cam (or wedge) to their released positions, thereby releasing the brakes. The clevis or yoke threads onto the brake chamber pushrod, which is also threaded to receive it.

Task A2.3 Inspect, adjust, repair, or replace manual and automatic slack adjusters.

The slack adjuster is critical in maintaining the required free play and adjustment angle. Current slack adjusters are required to be automatically adjusting, but their operation must be verified routinely.

Slack adjusters connect the brake chambers with the foundation assemblies on each wheel. They are connected to the pushrod of the brake chamber by means of a clevis yoke (threaded to the pushrod) and pin. Slack adjusters are spline-mounted to the S-cams and positioned by shims and an external snap ring.

The slack adjuster converts the linear force of the brake chamber rod into rotary force or torque and multiplies it. The distance between the slack adjuster S-cam axis and the clevis pin axis will define the leverage factor; the greater this distance, the greater the leverage.

The objective of brake adjustment, whether manual or automatic, is to maintain a specified drum to lining clearance and a specified amount of free play. Free play is the amount of slack adjuster stroke that occurs before the linings contact the drum.

Slack adjusters are lubricated by grease or automatic lubing systems. The seals in slack adjuster are always installed with the lip angle facing outward; when grease is pumped into the slack adjuster, grease will easily exit the seal lip when the internal lubrication circuit has been charged.

Automatic slack adjusters may require periodic adjustment. The method varies by manufacturer. Manual slack adjusters are adjusted by rotating the adjusting screw until the shoes are forced into the drums, then backing off the adjusting screw to set the specified free play. When a wheel-up adjustment is performed, free play is set to minimum drag of the shoe/drum relationship.

When replacing a failed slack adjuster, the distance between the axis of the S-cam bore and the clevis pin bore must be maintained; a difference of one-half inch can greatly alter the brake torque and unbalance the brakes.

Task A2.4 Inspect or replace cams, rollers, shafts, bushing, seals, spacers, retainers, brake spiders, shields, anchor pins, and springs.

When performing foundation brake servicing, all of the hardware mounted on the axle spider should be inspected and replaced if required. There should be no radial play of the S-camshaft, rollers and cams should be inspected for flat spots, and it is good practice to replace any spring steel components such as retraction springs and snap rings.

Spider fastener integrity should be checked at each brake job.

Spiders should be inspected for cracks at each brake job.

S-cam bushing seals should be installed so that the spider external bushing sealing lip faces inboard and the internal bushing sealing lip faces outboard from the bushing; this allows excess grease to exit from the inboard side of the bushing and prevents it from being pumped into the foundation assembly.

S-cam profiles depend on friction and should never be lubricated.

Task A2.5 Inspect, adjust, repair, or replace wedge-type brake housing, plungers, and wedge assembly.

Drum brakes can be classified as either cam-actuated or wedge-actuated. Wedge brakes can be actuated on two planes and these have higher theoretical braking efficiencies than drum brakes. They are used today in certain applications, often as front brakes on a highway tractor.

Wedge brakes are actuated by a brake chamber assembly: the brake chamber pushrod drives a wedge between a pair of plungers that force the shoes into the drums.

Wedge-type braking systems are self-adjusting and thus require fewer manual adjustments; however, they are vulnerable to seizure when equipment is not regularly used.

A specified lining-to-drum clearance must be maintained and this is checked with thickness gauge inserted at either end of each shoe; the drum to lining clearance should be consistent in all four measuring locations.

Task A2.6 Inspect, repair, or replace wedge brake spiders, manual adjuster plungers, and automatic adjuster plungers.

Single-actuated wedge brakes use a single plunger housing mounted on the spider assembly. These use shoes with fixed anchors, mounted to the spider.

Double-actuated wedge brakes use a pair of floating shoes actuated by two sets of plunger housings and brake chambers. The shoes float and are retained to the spider by clips and to the plunger housings by retraction springs that link each shoe.

There is a manual adjusting mechanism located on each plunger assembly consisting of a star wheel, which rotates a threaded plunger from the housing either inward or outward. The automatic adjuster consists of a spring-loaded pawl that rides against helical grooves on the exterior of the plunger actuator; as the brake friction linings wear, pawl ridges ratchet to a new position on the plunger actuator thus limiting retraction travel and defining free play.

When the plunger assemblies seize, the automatic adjusters cease to operate and the plunger assembly must be serviced. Plunger assemblies must be lubricated with high temperature lubricant on reassembly.

Task A2.7 Inspect, clean, rebuild/replace, and adjust air disc brake caliper assemblies.

Air-actuated heavy duty truck disc brakes have some similarities to automobile hydraulically actuated disc brakes. Air disc brakes use an air chamber and pushrod to apply braking torque to a powershaft. The powershaft has an external helical gear that acts on an actuator nut to create the clamping force required by the caliper to effect retarding effort on a rotor that turns with the wheel assembly. In other words, axial force applied to the powershaft is converted to clamping force by the caliper assembly.

Within the caliper assembly, pistons transmit the clamping force to brake pads which effect retarding effort by converting the kinetic energy (energy of motion) to friction and then heat.

Short outboard friction pad service life is usually an indication of caliper assembly seizure, that is, the sliding action of the brake caliper ceases. The caliper slide pins are usually at fault.

The rotors used on truck air disc brakes are usually vented to aid in the dissipation of brake heat. In vocational applications which require constant use of the brakes, air disc brake rotors have been vulnerable to heat related failures such as warpage, which has meant that they are not commonly used.

Task A2.8 Inspect and replace brake shoes, linings, or pads; determine needed repair.

Truck brake shoes are in most cases fixed anchor assemblies mounted to the axle spider and actuated by an S-camshaft. Brake shoes for a heavy duty truck can have their friction facings or linings mounted to the shoe by bonding, riveting, or by fasteners. In most current applications, the shoes are remanufactured and replaced as an assembly, that is, with the new friction facing already installed. When reusing shoes, they must be checked for arc deformities usually caused by prolonged operation with out-of-adjustment brakes. The lining blocks are tapered and seldom require machine arcing.

The brake shoes must be fitted with the correct linings. Lining requirements can change with different vehicles and different applications. When reconditioning shoes with bolted linings, if the original fasteners are reused, the lock washers should at least be replaced. Riveted linings should be riveted in the manufacturer's recommended sequence.

The friction rating of linings is coded by letter codes. Combination lining sets of shoes are occasionally used; these use different friction ratings on the primary and secondary shoes. When combination friction lining sets are used, care should be taken to install the lining blocks in the correct locations on the brake shoes.

It is good practice to replace the brake linings on all four wheels of a tandem drive axle truck on a preventive maintenance inspection (PMI) schedule. When the linings on a single wheel are damaged such as in the event of wet seal axle lube failure, the linings of both wheel assemblies on the axle should be replaced to maintain brake balance.

The friction pads in air disc brake assemblies should also be changed in paired sets, that is, both wheels on an axle. Vented disc brakes may have thicker inner pads than outer pads, while solid disc brakes usually have equal thickness in the inner and outer pads. Heat transfer is generally more uniform in the solid disc assemblies than vented ones, requiring thicker inboard pads.

Task A2.9 Inspect, reface, or replace brake drums or rotors.

If brake drums are to be reused (recent practice and the low cost of drums usually results in drum replacement with a brake job), they should be inspected for size specification, heat discoloration, scoring, glazing, threading, concaving convexing, bellmouthing, and heat checking. Used truck brake drums can seldom be successfully machined due to heat tempering. It is good practice to machine new drums before installation because of warpage caused by incorrect storage practices after manufacture.

The maximum legal service limit for 14-inch diameter brake drums is 0.120 inches. The machining limit is 0.060 inch and the discard limit is 0.090 inch. A drum micrometer is used to check brake drum diameter.

Brake rotors must be visually inspected for heat checking, scoring, and cracks. Rotors must be measured for thickness with a micrometer. A dial indicator is used to check rotor runout and parallelism. In highway applications, rotors tend to outlast drums and are often reused after a brake job. They must be turned within legal service specifications using a heavy duty rotor lathe.

3. Parking Brakes (5 Questions)

Task A3.1 Inspect and test parking (spring) brake chamber, diaphragm, and seals; replace parking (spring) brake chamber.

FMVSS 121 requires that all trucks be equipped with a mechanical parking and emergency brake system. The parking brake system is generally a separate system from the service brake system, with its own lines, chambers, and valves. The mechanical braking force is obtained by springs located in spring brake chambers. These use air acting on a holdoff diaphragm to release the spring brake.

The spring in a spring brake chamber requires approximately 60 psi acting on its holdoff chamber to fully release it. It is therefore capable of applying the equivalent of a 60 psi air application when there is no air acting on the holdoff chamber.

Spring brake chambers are dual chamber assemblies that combine a service application chamber in the front section and a spring chamber in their rear section. Their size rating (such as 30/30, 24/24, 30/24) refer to the surface area of the service and holdoff diaphragms. The caging port at the rear of the spring diaphragm is sealed with a plastic seal; if this seal is not in place, the internal components can rapidly corrode.

Spring brake chamber assemblies fail when either the service or holdoff diaphragms puncture or when the spring breaks. Spring breakage is often indicated by an air leak at the holdoff diaphragm that occurs when the fractured spring punctures it. It is essential that spring brakes be handled with caution even when it is known that the main spring has fractured.

Task A3.2 Inspect, test, or replace parking (spring) brake check valves, lines, hoses, and fittings.

The supply of air to the parking brake circuit is pressure-protected so that in the event of a complete primary or secondary circuit failure, there is sufficient air for a complete apply/release sequence. Restrictions in the lines supplying the holdoff chambers can slow both application and release times. Spring brake valve operation can be verified with an air pressure gauge. Lines supplying the holdoff chambers on each wheel should be of equal length; release timing is unbalanced if this is not the case.

Task A3.3 Inspect, test, or replace parking (spring) brake application and release valve.

Spring brake valves are often incorporated in multifunction valves. It is critically important that when valves malfunction, they be replaced by matching the part numbers, remembering that two valves that appear identical may have widely different performance characteristics.

The parking brake system of a tractor-trailer combination manages holdoff pressure delivered to the spring brake chambers. Two- and three-dash valve systems are used. FMVSS 121 requires that the shape of each valve be as follows (the color of each valve is an industry standard):

- Yellow diamond/square: the system spring brake (park) control valve. This valve masters the parking brakes on a tractor-trailer combination. It is pushed inwards to release the parking brakes. When pulled out to its park position, it cuts the air supply to the trailer (thus applying its park/emergency brakes) and puts the tractor into park/emergency mode.
- Red octagonal knob: the trailer supply valve. The system spring brake control valve must be pushed in an actuated before the trailer supply valve will work. This valve supplies system air to the trailer(s) and must be ON to release the trailer spring brakes.
- Blue round: isolates the tractor park brake function from that of the trailer. This valve is mastered by the spring brake control valve.

Task A3.4 Manually release and reset parking (spring) brakes.

Spring brake chambers are usually equipped with a cage bolt. The cage bolt is threaded and has a pair of lugs that engage to an internal cage plate. When a nut is turned down the cage bolt, the cage plate is pulled inwards, compressing the main spring of the spring brake chamber. This releases the parking brake. If the vehicle on-board air supply is available, a spring brake may be caged by chocking the wheels and releasing the parking brakes (supplying air to the holdoff chambers) and then installing the cage bolt to the cage plate with the main spring already compressed. It is important to ensure that the lugs in the cage bolt are properly engaged in the cage plate. Cage plates are manufactured out of aluminum alloy and those in older spring brake assemblies were susceptible to corrosion. Caged spring brake assemblies should always be handled with a great amount of care. Spring brake chambers should always be removed and installed fully caged.

In the past, failed holdoff diaphragms were routinely replaced. This is not current practice and when spring brake holdoff diaphragms fail, a piggy-back assembly (spring brake assembly minus the service chamber) or entire spring chamber is used to repair the condition. The spring brake chamber band clamps should never be removed with the main spring in uncaged condition. Many current spring brake assemblies have chamber clamps that cannot be unbolted.

It is illegal to discard spring brakes without first disarming them. Disarming a spring brake means releasing the main spring. This is achieved by placing the entire spring brake assembly in a disarmament chamber with the main spring uncaged, torch-cutting the spring brake chamber clamps, and separating the chamber. The entire assembly should remain in the disarmament chamber until the main spring can be observed to be free.

4. Antilock Brake System (ABS) (5 Questions)

Task A4.1 Observe antilock brake system (ABS) warning light operation; determine if further diagnosis is needed.

All current antilock brake systems (ABS) are electronically managed. A typical ABS system adapts a standard air or hydraulic system for electronic management. All current systems are designed to default to nonmanaged operation in the event of an electrical or electronic malfunction. Most new trucks are equipped with ABS brakes and these will become mandatory in the near future.

An ABS system consists of a means of monitoring the speed of each wheel, an electronic control unit (ECU) or ABS module to manage the system, and a modulator assembly to manage the outcomes of the ECU logic processing by modulating the service application pressures to each wheel. ABS brakes are especially effective at managing split coefficient braking conditions. Split coefficient braking occurs when road surface conditions (icy, wet, gravel, dry, etc.), differ from one wheel to another.

A fail relay coil is installed between the vehicle power source and the modulator valve assembly and is designed to illuminate a warning light in the cab in the event of a system failure. When this occurs the system fault must be diagnosed.

Task A4.2 Diagnose antilock brake system (ABS) electronic control(s) and components using self-diagnosis and/or recommended test equipment; determine needed repairs.

Although ABS systems are available in two-, four-, or six-channel designs, the four-channel system is commonly used today in tandem drive axle trucks. The four-channel system has a sensor on each of the steer-axle wheels, and on two of the four rear wheels, both on the same axle. The sensors input wheel speed data to the ABS module. The ABS module outputs an electrical signal to the four modulator valves. Two modulator valves control each wheel on the steer axle. The other two control service braking on each side of the tandem drive axles: one for the right side service chambers, the other for the left side.

The ABS wheel sensors use a reluctor wheel (serrations on a disc or the brake drum) and pickup to send a small AC electrical signal to the ABS module; the signal voltage value increases proportionally with wheel speed. In this way, a lockup condition can be sensed by the ABS module. The ABS module controls the system modulator valves. The modulator valves control the service application pressures to each of the front wheels and to each side of the tandem drive assembly. When a lockup condition is sensed by the ABS module, the air to the service chamber(s) controlled by the modulator can be momentarily relieved or pulsed. In current air brake systems, the pulsing speed of each modulator peaks at about four times per second.

A six-channel system on a tandem drive axle tractor monitors and modulates each of the six wheels on the vehicle. Two-channel systems monitor each side of one of the drive axles and modulate the service brakes on each side of the drive axle assembly by having each modulator valve actuate a pair of service chambers. Most trailers use a two-channel system.

All electronic ABS systems are equipped with self diagnostics. These can be accessed by flash or blink codes using the dash warning light or a digital diagnostic reader connected either at the American Trucking Association (ATA) connector in the cab or directly to the ABS module. All truck ABS systems use J1939 communication protocols, which means that one manufacturer's system can be read by any other.

The ABS system should first be accessed by selecting the correct message identifier (MID) and sequence through to identifying the failure mode indicator (FMI) that has triggered the fault code.

An Overview of the System

Task A4.3 Diagnose poor stopping and wheel lockup caused by failure of the antilock brake system; determine needed repairs.

When the ABS system malfunctions, the brake system reverts to functioning as any non-ABS managed brake system. When diagnosing a poor stopping and wheel lockup complaint, the technician should determine whether the problem is an electrical or electronic problem by scanning the ABS module. It should always be remembered that an air ABS system is still essentially and air brake system, and a modulator valve, while capable of being controlled electronically, is fundamentally a relay valve.

Task A4.4 Inspect, test, and service antilock brake system air, electrical, and mechanical components.

All current ABS systems manage subcomponent failures with safety-first failure strategy. Minor system malfunctions result in the illumination of the antilock condition lamp on the dash and the logging of a fault code by the ABS module. Depending upon the type of failure, the ECU may continue an ABS management at a lower performance level by disabling a portion of the ABS circuit or shutting the ABS management down, in which case the system reverts to standard air braking.

Manufacturers use sequential troubleshooting charts to troubleshoot ABS circuits and their components. It is performed each step in the sequence exactly as prescribed. The tools used to troubleshoot ABS systems are electronic service tools (scan tools, reader-programmers, PCs), digital volt ohmmeters, and proprietary connector and terminal assembly tools.

Task A4.5 Service, test, and adjust antilock brake system speed sensors following manufacturer's recommended procedures.

When a fault occurs in an ABS system, the ECU may shut down the malfunctioning portion of the circuit (that is, have it revert to standard air brake operation), but continue to manage the rest of the system in ABS mode. Wheel speed sensors use a reluctor wheel and pickup principle to produce an AC signal to the ECU. Wheel sensors will only function properly if the air gap (distance between the rotating serrated wheel and the stationary pickup) is properly set. Wheel sensor performance can also be affected by road dirt contamination and impact damage. Wheel sensor malfunction will produce a fault code in the ECU.

B. Hydraulic Brakes Diagnosis and Repair (18 Questions)

1. Hydraulic System (8 Questions)

Task B1.1 Diagnose poor stopping, pulling, premature wear, noise, or dragging complaints caused by hydraulic system problems; determine needed repairs.

The potential energy of a hydraulic brake system is mechanical force created by the action of a driver's foot acting on a brake pedal, usually assisted proportionally by pedal geometry leverage and a power assist system. In any hydraulic circuit, it can be assumed that the hydraulic medium is not compressible. If force is mechanically applied to a liquid in a closed system, it will be transmitted equally by the liquid to all parts of the system. Force applied by a master cylinder is transmitted equally throughout the hydraulic system though it may be modulated by valves in parts of that system.

The hydraulic circuit of a hydraulic brake system consists of a master cylinder, proportioning valves, metering valves, pressure differential valve and wheel cylinders. All truck hydraulic brake systems have dual circuits, meaning that in the event of a failure in one of the circuits, the other will back it up to effect at least one stop. As in truck air brake systems, the circuits are defined as the primary and secondary circuits. A failure of any moving part within the hydraulic circuit may cause pressure to become entrapped in a portion of the circuit, and this may cause dragging brakes or slow release times.

Task B1.2   Pressure test hydraulic system and inspect for leaks.

Pressure values within the hydraulic circuit may be tested with pressure gauges. The hydraulic system circuit can be pressurized simply by starting the vehicle engine and applying the brakes by foot pressure. External leaks may be verified by cleaning the externally visible portions of the circuit and applying the brakes. Internal leaks are more difficult to locate. Internal leakage within a master cylinder can be verified by using gauges plumbed to each portion of the hydraulic circuit.

Task B1.3   Check and adjust brake pedal pushrod length.

Most truck hydraulic brake systems use the brake pedal assembly to provide added leverage to the mechanical force provided by the driver's foot pressure. Brake pedal pushrod adjustment should always be made according to manufacturer's specifications.

Task B1.4   Inspect, test, or replace master cylinder.

The master cylinder converts the mechanical force applied to it by driver foot pressure and the brake booster system into hydraulic pressure to actuate the primary and secondary circuits of the brake system. It usually consists of integral reservoirs (one for each circuit), cylinder housing, compensating ports, return springs, and primary and secondary pistons. The primary piston is actuated mechanically. Pressure developed in the primary portion of the master cylinder charges the primary circuit (this can actuate either the front or rear brakes) and the secondary piston. When the master cylinder is operating normally, the secondary piston is actuated hydraulically, by whatever pressure value is developed in the primary portion of the cylinder, to charge the secondary circuit.

When the mechanical force applied to the primary piston is relieved, return springs acting on both the primary and secondary pistons return them to their original positions, permitting the fluid applied to each circuit to return to the reservoirs. Both sections of the master cylinder are aspirated with brake fluid by fill and compensating ports. Each piston is sealed in its bore by rubber seals. Primary and secondary circuit fluid do not come into contact with each other under normal operation.

When a failure occurs in either circuit, the pressure differential light in the dash will illuminate the first time the brakes are applied following the failure. If the failure occurs in the primary circuit, the primary piston will be forced through its travel without generating any fluid pressure until it contacts the secondary piston and mechanically actuates the secondary circuit. If the failure has occurred in the secondary circuit, the primary circuit will function normally, but the actuation of the secondary piston will result in no pressure delivered to the secondary circuit. In either case, the vehicle should be brought to an immediate standstill and not operated until a repair has been undertaken.

When testing a master cylinder in a hydraulic braking system, a liquid-filled hydraulic test gauge should be used. Deteriorated fluid, deteriorated seals, or a mixture of incompatible fluids may cause sludge and particulate that can plug fill and compensating ports, resulting in slow application times, slow release times, and brake failure.

Master cylinders in truck hydraulic systems are routinely reconditioned when seals and springs are replaced, cylinders honed, etc. When cleaning master cylinders before reassembly, only isopropyl alcohol should be used to clean components; use of solvents can swell the seals and leave behind corrosive residues.

Task B1.5 Inspect, test, or replace brake lines, flexible hoses, and fittings.

Steel tubing should be checked for wear, dents, kinks, and corrosion. Preflared and preformed tubing helps reduce custom cuts, bending, and flaring of new tubing. Two types of flaring styles and seats are used: International Standards Organization (ISO) and double flare. ISO uses an outward flare, while a double flare creates a double wall at the nipple seat for greater strength. When cleaning brake tubing, only isopropyl alcohol should be used because of its noncorrosive characteristics, ability to evaporate rapidly, and residue-free drying. Residues remain when using other cleaning agents such as soap and water, mineral spirits, and hydraulic brake fluid. When bending tubing, a tube bender should be used and the line filled with fine silica sand to prevent collapsing.

Task B1.6 Inspect, test, and replace metering (holdoff), load sensing/proportioning, proportioning, and combination valves.

The metering valve is used on vehicles equipped with front disc and rear drum brakes. It is required to achieve brake timing balance during light brake applications by withholding the delivery of application pressure to the front disc brakes until pressure exceeds a predetermined value in the circuit responsible for actuating the rear brakes. This lag or delay is required so that hydraulic pressure builds sufficiently in the rear brake hydraulic circuit to overcome the tension of the rear brake shoe return springs and the free travel of the shoes. The objective is to enable simultaneous application of both front and rear brakes. For this reason, the metering valve is sometimes known as a holdoff valve.

When a pressure bleeder is used to bleed any system equipped with a metering valve, the manufacturer's instructions as to how to open the valve must be observed. When manually bleeding a brake system, application of the brake pedal develops sufficient pressure to overcome the metering valve opening pressure.

A proportioning valve is also used on systems combining front disc and rear drums. The proportioning valve is installed in the circuit supplying the rear brakes. Its function is to reduce the application pressure to the rear wheel cylinders and prevent rear wheel lock-up. Disc brakes require higher hydraulic application pressures.

The proportioning valve operation should be verified at each brake inspection or if reported for a rear wheel lockup condition. To check valve operation, hydraulic gauges should be installed ahead and behind the valve or alternatively, to each of the two circuits.

A load or height-sensing valve is used on some systems to sense vehicle load transfer effect under braking; the valve proportions front and rear braking, correlating it to weight transfer during braking. The valve is located on the vehicle frame cross member and is activated through a linkage system connected to the rear axle housing.

Task B1.7 Inspect, test, repair, or replace brake pressure differential valve and warning light circuit switch, bulbs, wiring, and connectors.

The pressure differential valve is also known as a brake light warning valve and dash lamp valve. It consists of a cylinder through which primary and secondary hydraulic pressure act on either side of a spool.

When the pressure in both the primary and secondary circuits is equal, the spool floats in a neutral position. Should a pressure imbalance occur in either the primary or secondary circuit, the spool will shuttle to one side of the cylinder and in doing so, ground an electrical signal, illuminating a dash warning light.

The pressure differential valve will normally re-center automatically upon the first application of the brakes after repairs are completed. Some pressure differential valves will require manual resetting.

The proportioning valve should be inspected whenever the brakes are serviced. This inspection should include the electrical warning light circuit.

Task B1.8 Inspect, clean, and rebuild or replace wheel cylinders.

A wheel cylinder is the actuator of a hydraulic brake system; it is also known as a slave cylinder. It is supplied with hydraulic brake fluid supplied by the master cylinder, and converts hydraulic pressure to mechanical force at the foundation brake assembly. Wheel cylinders are most generally constructed of cast iron for higher durability and lower manufacturing costs.

Most wheel cylinders are double acting. They are machined out of cast iron and house two pistons within a cylinder bore. The pistons are sealed in the cylinder by rubber seals. When the pistons are subjected to hydraulic pressure, they are forced outward to actuate brake shoes, forcing them into the drum. Manual and automatic adjusting mechanisms are integral in the wheel cylinder assembly.

Wheel cylinders are commonly reconditioned. The cylinder bores can be honed and most of the internal components replaced if required. They are vulnerable to contaminates in the hydraulic fluid. When the cylinder bores become scored, the result is fluid leakage from the cylinder past the seals. Seals may fail if exposed to chemical contaminants. Each wheel cylinder is equipped with a bleed port. You crack open this bleed port to purge air from the hydraulic circuit in the bleeding process.

Task B1.9 Inspect, clean, and rebuild or replace disc brake caliper assemblies.

Fixed and sliding or floating calipers are used in hydraulic brake systems. In a fixed-type caliper, the rotor sits in between two or four pistons, and the clamping action of opposed pistons acting on the rotor is responsible for retarding the rotor. In a floating or sliding caliper, one or more pistons sit on one side of the rotor, and the caliper housing is designed to slide when the piston is subjected to hydraulic pressure; this action effects the clamping action of the rotor. The friction surfaces that contact the rotor are brake pads. Disc brakes are not self-energizing and in general greater hydraulic pressures are required to apply them. They operate at higher mechanical efficiencies than drum brakes. Disc brake calipers are reconditioned in the same manner as wheel cylinders. Manufacturer's tolerance specifications must be observed during reconditioning.

Task B1.10 Inspect/test brake fluid; bleed and/or flush system.

It is good practice to replace the system brake fluid at each major brake overhaul, especially if the failure has been caused by seal failure or fluid contamination. Using the manufacturer's specified brake fluid is important to ensure proper operation of the system. The approved heavy duty brake fluid should retain the proper consistency at all operating temperatures. It will not damage rubber cups and helps to protect the metal parts of the brake system against failure. Water, mineral spirits, and gasoline should never be used to flush a hydraulic braking system because of incompatibility with other materials and corrosive effects. Alcohol or compatible brake fluid is always used to flush hydraulic braking systems, and these fluids should not be reused after the flushing is complete.

A container storing brake fluid must always be tightly sealed when not in use to prevent moisture from being absorbed into it. Mineral oil, alcohol, antifreeze, cleaning solvents, and water, even in very small quantities, will contaminate most brake fluids. Brake fluid has a shelf life of 1 year. This shelf life diminishes if the storage container is not completely full. Mixing of different brake fluid can cause coagulation and result in hydraulic failure; it is important to ensure that fluids are compatible when topping up system reservoirs.

Brake systems may be bled using a bleeder ball or manually. The first method is preferred as the operation can be performed by one person. In all cases, system air is purged from the valves and wheel cylinders, in a sequence defined by the manufacturer.

Task B1.11 **Inspect, test, repair, or replace hydraulically controlled parking brake components and systems.**

Some truck hydraulic brake systems are equipped with a hydraulic parking brake system. These use a dedicated hydraulic circuit and have a parking brake cylinder located in the foundation assembly on the opposite side of the shoe to the wheel cylinder. The parking brake is actuated by a cab-located foot or hand lever.

2. Mechanical System (6 Questions)

Task B2.1 **Diagnose poor stopping, brake noise, premature wear, pulling, grabbing, dragging, or pedal feel complaints caused by drum and disc brake mechanical assembly problems; determine needed repairs.**

Hydraulic foundation brakes may use servo and nonservo principles. Servo action occurs when the action of one shoe is guided by the movement of the other, permitting both shoes to act as a single unit. Self-energization occurs when the shoes are driven into the drum and rotate fractionally as a pair with the drum, before grabbing.

Nonservo action is also a term generally used for certain drum/shoe-type brakes. Some original equipment manufacturers (OEMs) refer to these brakes as leading trailing shoe brakes. Nonservo brakes have each shoe working independently of each other to stop the vehicle. They are separately anchored. When actuated by the wheel cylinder, the shoe pivots on the anchor and is forced against the drum.

Disc brakes are nonenergized, and they require more force to achieve the same braking effort as self-energized drum brakes. However, they have superior mechanical efficiency and are often used on the front axle brakes of straight trucks; front-wheel brakes perform a higher percentage of braking on straight trucks because of their load transfer and suspension effects.

Grabbing and pulling can occur in the foundation brake assembly caused by broken hardware components, especially return springs, drum failures, malfunctions in the adjusting mechanism, and parking brake-related problems.

A primary cause of a pulsating pedal condition is warped disc brake rotors. Rotors warp due to overheating. Underspecifying the braking requirements of a vehicle or machining rotors too thin at overhaul can cause pedal pulsation and rotor warp. Thickness variations on disc brake rotors on different wheels can also cause pedal pulsation and loss of braking power.

Task B2.2 **Inspect, reface, or replace brake drums or rotors.**

Brake drums may reused if they are within the manufacturer's specifications. The critical specifications are the maximum wear limit, machine limit, and maximum permissible diameter. Drums are measured with a drum gauge and should be checked for out of round, bell-mouthing, convexing, concaving, and taper. Drums should be inspected after measuring for heat checks and cracks and before they are machined.

Disc brake rotors may be reused if they are within the manufacturer's specifications. They should be measured for thickness with a micrometer and checked for parallelism and run-out with a dial indicator. If within machine limits, the rotor may be turned on a rotor lathe.

Task B2.3 Inspect, adjust, or replace drum brake shoes/linings, mounting hardware, adjuster mechanisms, and backing plates.

When a brake job is performed, the brake shoes, return springs, and fastening hardware are replaced. The friction face codes should be observed when replacing brake shoes. Most brake shoes today use bonded friction blocks, but riveted and bolted types are still in existence. Ensure that primary (leading) and secondary shoes are installed in their correct locations. Brake shoes are the first component to wear out in duo-servo brake systems.

Task B2.4 Inspect, service, or replace disc brake pads, hardware, and mounts.

Most disc brake assemblies today have wear indicators which produce a noise when the wear limit is exceeded. Servicing disc brake pads usually involves removing the caliper assembly, ensuring that float pins are not seized, backing off the automatic adjusting mechanism, and installing a new pair of brake pads. When a self-adjusting mechanism is used, care should be taken to ensure it is properly activated on reassembly. Whenever the brake pads are replaced, the brake rotors should be both measured and visually inspected.

Task B2.5 Inspect, adjust, repair, or replace drive line parking brake drums, rotors, bands, shoes, mounting hardware, and adjusters.

Cable-actuated, drive line parking brakes are used in some hydraulic truck brake systems especially in air-over-hydraulic brake systems. The unit is a band and rotor assembly mounted at the rear of the transmission; when engaged, the rear wheels are locked stationary by means of the driveshaft.

Task B2.6 Inspect, adjust, repair, or replace drive line parking brake application system pedal, cables, linkage, levers, pivots, and springs.

A drive line parking brake is incorporated as part of the drive shaft and is mechanically controlled by means of a cable actuated from the cab. The parking brake uses a lever mechanism mounted within the foundation assembly to force the brake shoes into the drums without any assist from the wheel cylinders.

Because the parking brake is actuated by a cable extending the length of the vehicle, it is vulnerable to seizure, especially when not used for extended periods. The parking brake cable may be freed up using penetrating oil, but it more often requires replacement. All highway vehicles are legally required to have functioning parking brakes.

3. Power Assist Units and Miscellaneous (4 Questions)

Task B3.1 Diagnose poor stopping complaints caused by power brake booster problems; determine needed repairs.

In truck hydraulic brake systems, the use of a vacuum or hydraulically assisted brake booster is required to reduce the foot effort that must be applied to the master cylinder to actuate the brakes. Vacuum assisted boosters are not found on today's medium duty trucks using hydraulic brakes.

A hydraulic booster mechanism section is powered by either the truck's power steering pump or by a dedicated pump. The unit comprises an open center valve, reaction feedback mechanism, a large diameter boost or power piston, a reserve electric-powered pump, an integral flow control switch, and a power steering gear, operating in series. Vehicles with manual steering gear must use hydraulic boosters with a dedicated hydraulic pump.

Power brake boosters operate by assisting brake pedal effort with hydraulic pressure in proportion to pedal travel. Both single and dual diaphragm units are used. Malfunctioning hydro-boost units in trucks will usually require greatly increased brake pedal effort. The brake system cannot be effectively operated in this condition. Booster units may be repaired by overhaul or replacement.

Task B3.2 Inspect, test, repair, or replace power brake booster, hoses, and control valves.

In a hydromax power brake booster, if flow from the hydraulic pump is interrupted, an electric pump backs up the system. When replacing hydraulic hoses in the boost system, the lines must conform to the Society of Automotive Engineers (SAE) J189 standard.

Hydro-boost systems are bled by cranking the engine over without starting it (the ignition system should be disabled) with reservoir filled with the recommended fluid. The refilling procedure may have to be repeated.

Task B3.3 Test, adjust, and replace brake stop light switch, bulbs, wiring, and connectors.

Stoplights on hydraulic brake circuits may be actuated electromechanically or electrohydraulically. In either case, the switch simply grounds an electrical signal which closes the brake light circuit. The brake light circuit may be tested with a DVOM.

C. Wheel Bearings Diagnosis and Repair (4 Questions)

Task C1 Remove and replace axle wheel assembly.

The wheel assembly should be removed using a wheel dolly. The wheel seals should be removed using a heal bar or drift and a hammer. It is good practice to replace wheel seals each time they are removed from the hub. Unitized seals require replacement when removed because the rubber sealing surface is damaged on removal. Removing the wheel seal enables the bearing cone to be removed for cleaning and inspection; the cup may be inspected in the hub. If the bearing assembly has to be replaced, the cups may be driven out of their bores using an appropriately sized bearing driver. Alternatively, a mild steel drift and hammer may be used. The procedure is reversed to install cups. Never use a brass drift to install bearing cups.

Hubs should be prelubed when installing the wheel assembly. On nondriven axles, the bearing and hub assembly must be filled to a prescribed level in the calibrated inspection cover. In drive axles, the bearing and hub assembly is supplied with lubricant from the differential carrier.

Task C2 Clean, inspect, lubricate, or replace wheel bearings; replace seals and seal wear rings.

Wheel bearings should be inspected at each brake job and at each PM service that requires that the wheels be removed. Most current trucks use wet bearings; that is, they are lubricated with liquid lubricant, usually gear lube.

The use of taper roller bearings has become almost universal. A tapered wheel bearing assembly consists of a cone assembly and a cup or race. The cone assembly consists of tapered rollers mounted in a roller cage. The bearing race or cup is interference-fit to the hub. A pair of taper roller bearings support the load in each wheel hub. The cones and cups of taper roller bearings are not interchangeable; when damage is evident in either the cup or the cone, both must be replaced.

Bearings should be cleaned with solvent and air dried, ensuring that the cone is not spun out by the compressed air. Both the cone (rollers) and the cone should be inspected for spalling, galling, scoring, heat discoloration, and any sign of hard surface failure.

When reinstalling wet bearings, they should be prelubed with the same oil to be used in the axle hub. When grease-packed bearings are used, the bearing cone must be packed with grease. This procedure may be performed by hand or by using a grease gun and cone packer. Grease-lubricated wheel bearings require the use of a high temperature axle grease. It is bad practice to pack wet bearings with axle grease and it may shorten bearing life by reducing lubrication efficiency.

Bearings fail primarily because of dirt contamination. This may be due to the conditions a vehicle is operated under or poor service practice. Lubrication failures caused by an inappropriate lubricant or lack of lubricant (caused by a failed wheel seal) also account for a large number of bearing failures: lack of lubricant will result in rapid failure and a bearing welded to an axle.

Many bearing failures are caused by brinelling. Brinelling appears as indentations across the bearing race. This condition occurs when the bearing is not rotating coming from impact loading or vibration.

Maladjusted bearings can fail rapidly, especially when the preload is high. The result of high preload on a bearing can cause the bearing to friction-weld to the axle. Overloaded bearings fail over time rather than rapidly.

Task C3 Adjust axle wheel bearings in accordance with manufacturer's procedures.

Most highway heavy duty trucks use taper roller wheel bearings. The correct method of adjusting a taper roller wheel bearing is outlined by the Truck Maintenance Council (TMC) division of the American Trucking Association. All bearing manufacturers in North America have endorsed this method, which supercedes any previous adjustment procedures outlined by them. The TMC bearing adjustment method requires that the bearing be seated to a specified torque value at the adjusting nut while rotating the wheel bearing; this is designed to seat the bearing with a preload. Next, the adjusting should be backed off to between one sixth and one third of a turn to locate the adjusting nut to the jam mechanism. Finally, the bearing endplay must be measured with a dial indicator: the required specification is between 0.001 inch and 0.005 inch. Endplay must be present. If at least 0.001 inch endplay is not present, the adjustment procedure should be repeated.

The bearing setting is locked in place on the axle spindle by a lock or jam nut that should be torqued to specification after the adjustment procedure. In some axles, a split forged or castellated nut and cotter pin are used to retain the wheel assembly. Current recommended practice in the truck service industry is defined by the TMC, and the above method was developed to establish a single standard to counter numerous highway wheel-off incidents.

Sample Test for Practice

Sample Test

Please note the letter and number in parentheses following each question. They match the overview in section 4 that discusses the relevant subject matter. You may want to refer to the overview using this cross-referencing key to help with questions posing problems for you.

1. A truck is in for parking brake service. Technician A says the spring brake chambers should be "caged" before disconnecting any air line or hose. Technician B recommends that draining the service air system be one of the first operations in servicing truck parking brakes. Who is right?
   A. A only
   B. B only
   C. Both A and B
   D. Neither A nor B

   (A3.1 and A3.4)

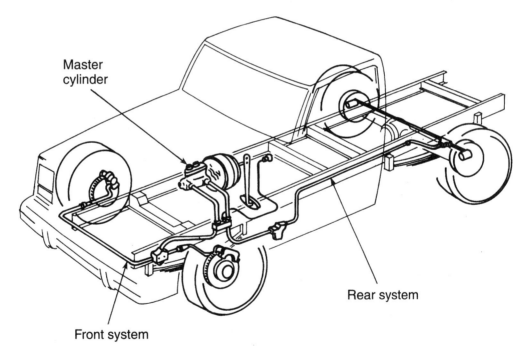

2. In the above figure, which of the following is part of a truck's hydraulic brake system?
   A. Slack adjuster
   B. Combination valve
   C. Treadle valve
   D. Tractor protection valve

   (B1.6)

3. Technician A says that the ABS ECU uses a series of light emitting diodes (LEDs) in the diagnostic window to indicate fault codes in the ECU. Technician B says that the number of LEDs in the diagnostic window vary depending on the system.
   A. A only
   B. B only
   C. Both A and B
   D. Neither A nor B   (A4.1)

4. All of the following are part of a heavy duty truck's foundation brakes **EXCEPT:**
   A. slack adjusters.
   B. hydraulic lines.
   C. brake pads.
   D. brake chambers.   (A2.2)

5. What does a bobtail proportioning valve do?
   A. It speeds up the exhaust of air from the air chambers.
   B. It speeds up the application and release of the brakes.
   C. It works with the park control valve to provide tractor protection and control trailer parking brake application.
   D. It automatically reduces the amount of air pressure that can be applied to the tractor drive axle(s).   (A1.17)

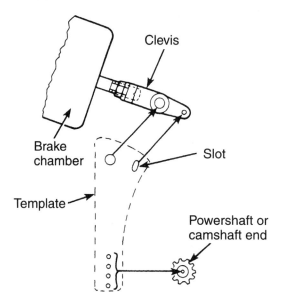

6. In the figure above, the clevis is being reinstalled onto its pushrod. Technician A says that the clevis should be correctly positioned on the pushrod in order to facilitate proper brake adjustment. Technician B says that the clevis on a brake chamber pushrod is keyed. Who is right?
   A. A only
   B. B only
   C. Both A and B
   D. Neither A nor B   (A2.2)

7. Technician A says that it is a good practice to change the brake fluid whenever you perform a major brake repair. Technician B says you should immediately replace contaminated brake fluid. Who is right?
   A. A only
   B. B only
   C. Both A and B
   D. Neither A nor B   (B1.10)

8. All of the following are part of an air compressor **EXCEPT:**
   A. piston rings.
   B. crankshaft.
   C. discharge valve.
   D. roller bearing. (A1.6)

9. A driver complains of insufficient service brake application when the service brake pedal is depressed. Which of the following is the LEAST likely cause?
   A. Restricted air flow or low air pressure to the service brake chamber
   B. Improper adjustment of slack adjuster and chamber pushrod
   C. Grease on shoe linings/faces
   D. A ruptured diaphragm (A1.1)

10. The first part to wear out and need replacing on a duo-servo hydraulic braking system is:
    A. wheel cylinder.
    B. drum shoes.
    C. anchor pins.
    D. return springs. (B2.3)

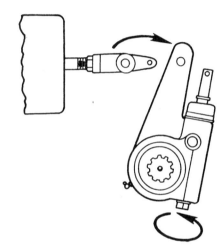

11. In the figure above, a slack adjuster should be adjusted so that the brake chamber pushrod is:
    A. less than 90° from the slack adjuster arm when fully applied.
    B. parallel to the slack adjuster arm when the brakes are fully applied.
    C. more than 90° from the slack adjuster arm when fully applied.
    D. at a 90° angle from the slack adjuster arm when fully applied. (A2.2)

12. What is the maximum number of reservoirs installed on a tractor-trailer configuration?
    A. Three
    B. Six
    C. Nine
    D. Twelve (A1.3)

13. Technician A says brinelling appears as indentations across the bearing raceway. Technician B says brinelling occurs when the bearing is not rotating. Who is correct?
    A. A only
    B. B only
    C. Both A and B
    D. Neither A nor B (C2)

14. All of the following could cause a service brake to apply or release sluggishly **EXCEPT:**
    A. misaligned linkage.
    B. defective chamber return spring.
    C. broken or weak brake return springs.
    D. obstructed chamber. (A2.1)

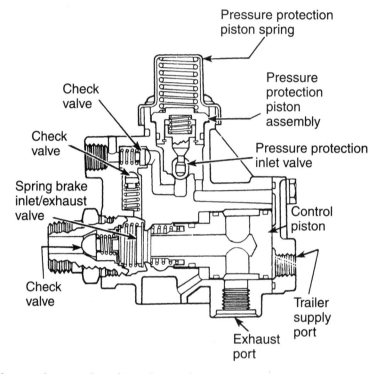

15. In the figure above, what does the trailer spring brake valve do?
    A. Releases air in the reservoir to apply the service brake.
    B. Allows the buildup of air pressure to allow operation of spring brake system.
    C. Redirects air to the spring brake chambers for initial application.
    D. Releases the trailer parking brakes with supply air pressure above 85 psi. (A1.19)

16. A technician has just replaced a wheel cylinder. Technician A says that the pressure differential valve will require manual resetting. Technician B says that after manually resetting the pressure differential valve, you should replace the proportioning valve. Who is right?
    A. A only
    B. B only
    C. Both A and B
    D. Neither A nor B (B1.7)

17. All of the following are parts of a slack adjuster **EXCEPT:**
    A. worm gear.
    B. pushrod.
    C. actuator piston.
    D. straight or offset clevis. (B2.3)

18. Technician A says you install the fail relay coil on an ABS system between the modulator assembly and a good ground. Technician B says power for the fail light flows through the normally closed relay contact(s). Who is right?
    A. A only
    B. B only
    C. Both A and B
    D. Neither A nor B (A4.4)

19. All of the following are possible drum conditions **EXCEPT:**
    A. overtorqued.
    B. concave.
    C. scored.
    D. threaded. (A2.9)

20. A heavy duty truck's brake shoe lining can be:
    A. integral.
    B. molded.
    C. staked.
    D. riveted. (A2.8)

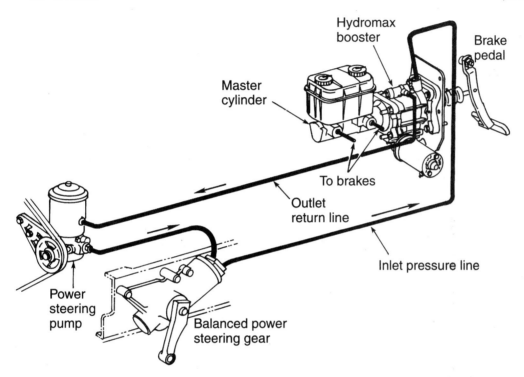

21. As shown in the figure above, technician A says that a hydraulic hydrobooster and balanced steering gear can be integrated with the power steering pump. Technician B says that a dedicated hydraulic power source not requiring a balanced steering gear can be used to power a hydraulic hydro booster.
    Who is right?
    A. A only
    B. B only
    C. Both A and B
    D. Neither A nor B (B3.2)

22. A governor cut-out test is being performed. Technician A says the governor should be visually examined. Technician B says to drain all reservoirs to 0 psi.
    Who is right?
    A. A only
    B. B only
    C. Both A and B
    D. Neither A nor B (A1.2)

23. On a full-floating axle, what supports the vehicle's weight?
    A. The axle housing
    B. The axle shaft
    C. The wheel assembly
    D. The inner axle bearing (C1)

Left wheel rotation

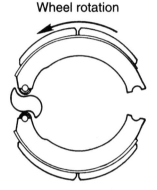
Wheel rotation

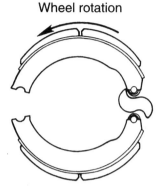
Wheel rotation

24. When replacing brake shoes on a tandem axle in a combination vehicle, the primary shoe for the left front tandem (with the cam in front of the axle) is positioned at what point in the figure above:
    A. to the left
    B. to the right
    C. on the top
    D. on the bottom (A2.8)

25. Technician A uses double flare tubing when replacing brake lines. Technician B uses ISO tubing when replacing brake lines. Who is right?
    A. A only
    B. B only
    C. Both A and B
    D. Neither A nor B (B1.5)

26. What is the LEAST likely cause of a noisy air compressor?
    A. A loose drive pulley
    B. Restrictions in cylinder head or discharge line
    C. A defective head gasket
    D. Inadequate lubrication of the unit (A1.5)

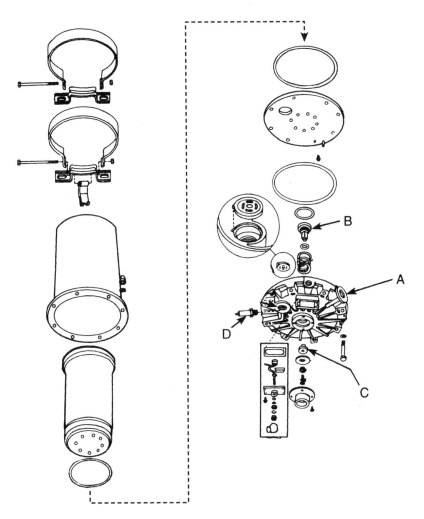

27. What is the part as shown in the figure above, marked "C"?
    A. Unloader valve
    B. Purge valve
    C. Solenoid valve
    D. Desiccant holder (A1.9)

28. When testing the operation of the trailer control valve, a technician should:
    A. move the handle to the fully applied position and record the air pressure.
    B. drain the air system to 0 psi, and then allow pressure to build up again.
    C. record the gauge reading of the air pressure first, and then drain the system.
    D. listen for air leakage around the handle as the handle is moved to the applied position. (A1.14)

29. What is the LEAST likely cause of poor braking power?
    A. Grease- or oil-contaminated shoes/linings
    B. Leaking quick-release valves
    C. Misadjusted slack adjuster/pushrod
    D. Leaking power booster (B1.1)

30. An ideal air braking system can be defined as:
    A. one in which braking forces are proportional on all axles.
    B. one in which the braking pressure reaches each actuator simultaneously and at the same pressure level.
    C. one in which each axle receives braking force in a sequential manner, starting with the closest axle to the governor.
    D. one in which each wheel receives a graduated braking force over a preset range of the most probable braking conditions. (A1.1)

31. All of the following could cause excessive leakage in an air brake system (service brakes applied) **EXCEPT:**
    A. a leaking brake chamber diaphragm.
    B. a leaking hose, tube, or fittings.
    C. defective gaskets.
    D. a bad relay valve. (A1.2)

32. The pushrod stroke is being measured. Technician A says that the applied stroke uses an 80-psi brake application. Technician B says the brakes are adjusted (if needed) to achieve the proper free stroke. Who is right?
    A. A only
    B. B only
    C. Both A and B
    D. Neither A nor B (A2.2)

33. Technician A adjusts wheel bearings while turning the wheel assembly. Technician B refers to the manufacturer's procedures when adjusting wheel bearings. Who is right?
    A. A only
    B. B only
    C. Both A and B
    D. Neither A nor B (C3)

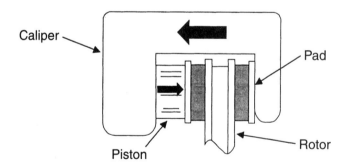

34. On an air disc brake in the figure above, what causes short outboard lining life?
    A. A piston seized in its bore
    B. A damaged rotor
    C. A damaged or defective brake hose/line
    D. A caliper seized on the slide pins (A2.7)

35. You are servicing an ABS system. Technician A says that wheel and axle sensors are electromagnetic devices used to obtain vehicle speed information. Technician B says that when an ABS system is disabled, the vehicle goes into its standard braking mode. Who is right?
    A. A only
    B. B only
    C. Both A and B
    D. Neither A nor B (A4.2)

36. A slack adjuster is being adjusted. Technician A says excessive wear limits between the clevis and the collar will cause the stroke to be too long. Technician B uses a spanner wrench to adjust an automatic slack adjuster. Who is right?
    A. A only
    B. B only
    C. Both A and B
    D. Neither A nor B (A2.3)

37. What is the LEAST likely result of using the wrong brake fluid?
    A. Damage to the cup seals of the wheel cylinder
    B. Absorption of moisture from the air
    C. Brake fluid boiling
    D. Contamination of the air in air over hydraulic systems (B1.2 and B1.10)

38. What should you not do, when replacing the diaphragm on a spring brake chamber?
    A. Slide the chamber sideways off the adapter.
    B. Rotate the chamber to break the diaphragm loose.
    C. Loosen the clamps by lightly striking them with a hammer or soft mallet.
    D. Exhaust the air pressure. (A2.2)

39. An aftercooler air dryer:
    A. requires periodic changing of its oil filter.
    B. employs no substances or materials to remove moisture.
    C. receives cool air from the compressor, and then removes the moisture from it.
    D. should be drained daily. (A1.10)

40. Where is a quick-release valve mounted?
    A. Close to the brake chambers
    B. In the cab, close to the trailer protection valve
    C. In the middle of a combination vehicle's axles for easier line routing
    D. On the outside backwall of the tractor next to the gladhands (A1.16)

41. A Rotochamber is being serviced. Technician A says that a type 30 Rotochamber has an effective diaphragm area of 30 square inches. Technician B says a Rotochamber's operation is similar to the conventional type brake chamber. Who is right?
    A. A only
    B. B only
    C. Both A and B
    D. Neither A nor B (B2.3)

42. Technician A says it's okay to machine oversize drums as long as the drum's manufacturer's recommendation for machining dimensions are followed. Technician B says that turning an oversize drum does not sacrifice its strength. Who is right?
    A. A only
    B. B only
    C. Both A and B
    D. Neither A nor B (B2.2)

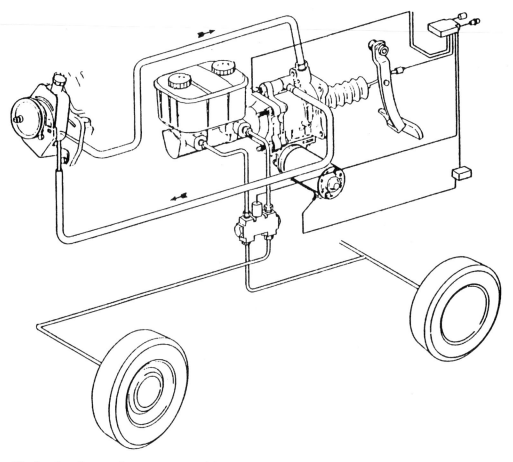

43. In the figure above, a typical hydraulic power booster interfaces with the following electrical and mechanical components **EXCEPT:**
    A. a buzzer.
    B. drum/drum, disc/drum, or disc/disc foundation brakes.
    C. the vehicle's storage batteries.
    D. the alternator. (B3.3)

44. A wheel cylinder:
    A. does not need any maintenance.
    B. needs to be inspected at a PM.
    C. needs to be replaced periodically.
    D. can be constructed of aluminum. (B1.8)

45. What happens when a failure is detected in an ABS system?
    A. The ECU will no longer cover that portion of the system.
    B. Full function ABS will still be available.
    C. The ECU will completely shut down.
    D. The ECU will delay implementing a default until approximately 5 minutes have elapsed. (A4.3)

46. Technician A says that if any slack adjuster defect is found, the slack adjuster is to be replaced. Technician B says that an automatic slack adjuster needs no lubrication, due to permanent lubrication. Who is right?
    A. A only
    B. B only
    C. Both A and B
    D. Neither A nor B (A2.3)

47. You are performing an air timing balance test on a combination vehicle. Technician A uses a stopwatch as part of the required equipment. Technician B says that balanced air timing means each axle receives air at the same time and at the same pressure level as needed by the driver. Who is right?
    A. A only
    B. B only
    C. Both A and B
    D. Neither A nor B (A1.2)

48. What is the difference between a fixed caliper and a sliding caliper on a truck with air disc brakes?
    A. Fixed calipers are less expensive than a sliding caliper.
    B. Sliding or floating calipers are less expensive than fixed calipers.
    C. Fixed caliper pads are smaller than sliding caliper pads.
    D. A fixed caliper has pistons on each side of a brake rotor, while a floating or sliding caliper only has piston(s) on one side of a brake rotor. (A2.7)

49. You are servicing a piggyback spring brake assembly. Technician A says no attempt should be made to repair/replace any part of the piggyback assembly, but that it should be completely replaced as a unit. Technician B says that the spring chamber should be disarmed (manually released), before discarding. Who is right?
    A. A only
    B. B only
    C. Both A and B
    D. Neither A nor B (A2.3)

50. The dual circuit brake application valve:
    A. has a separate inlet/exhaust valve that controls two separate reservoirs.
    B. is usually suspended from the cab firewall.
    C. is a push-pull valve mounted on the panel in the cab.
    D. operates in a clockwise manner and releases with a counterclockwise motion. (A1.11)

51. Technician A says that a servo-type brake uses a different primary and secondary shoe. Technician B says that the primary shoe in a servo-type brake system has a weaker return spring. Who is right?
    A. A only
    B. B only
    C. Both A and B
    D. Neither A nor B (B2.3)

52. What of the following conditions can cause brake pedal fade?
    A. A seized brake caliper piston
    B. Brake drum machined beyond its limit
    C. Leakage past the master cylinder cups
    D. Air in the hydraulic system (B1.1)

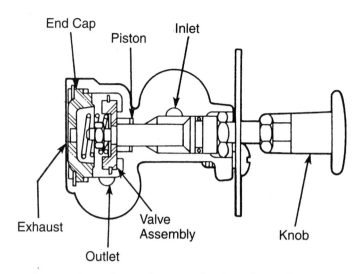

53. In the figure above, when the park control valve closes (button out):
    A. the spring brake releases.
    B. air begins to build up in the piggyback.
    C. it will not reopen until the treadle valve is depressed and opened.
    D. air in the rear portion of the spring brake chamber is exhausted. (A3.2)

54. Technician A says when wedge brakes are not properly adjusted, this reduces brake advantage and contributes to poor brake performance. Technician B says that a properly adjusted brake is one where the angle between the slack adjuster and the chamber pushrod is 90° when the brakes are applied. Who is right?
    A. A only
    B. B only
    C. Both A and B
    D. Neither A nor B (A2.6)

55. A truck with drum brakes is being serviced. Technician A says that the brakes may be wedge-actuated. Technician B says that the brakes may be cam-actuated. Who is right?
    A. A only
    B. B only
    C. Both A and B
    D. Neither A nor B (A2.5)

56. A holdoff valve is another name for:
    A. a pressure differential valve.
    B. a metering valve.
    C. a load proportioning valve.
    D. a combination valve.

57. A lining is being installed on its backing shoe. Technician A uses the same sequence of riveting as on the previous shoe. Technician B follows the manufacturer's procedures when riveting linings on shoes. Who is right?
    A. A only
    B. B only
    C. Both A and B
    D. Neither A nor B (B2.3)

58. Technician A says that when a front disc-rear drum hydraulic system is used, a gladhand valve is also used. Technician B says that only some OEMs use a proportioning valve in a front disc-rear drum non-ABS braking system. Who is right?
    A. A only
    B. B only
    C. Both A and B
    D. Neither A nor B (A1.21)

59. A driver communicates a complaint of dragging brakes. Technician A says a leaking diaphragm or seal could cause the problem. Technician B says a broken return spring in the spring brake section could be the cause. Who is right?
    A. A only
    B. B only
    C. Both A and B
    D. Neither A nor B (A1.1)

60. While road testing a truck, after a master cylinder overhaul, the truck brakes drag. Technician A says that the compensation ports may be blocked. Technician B says that excessive pedal free play could be the cause. Who is right?
    A. A only
    B. B only
    C. Both A and B
    D. Neither A nor B (B1.3)

61. If a technician fails to install master cylinder reservoir cover diaphragm, which of these will it cause?
    A. Fluid contamination
    B. Loss of pressure
    C. Seal erosion
    D. Air in the fluid (B1.4)

62. You are inspecting a four-wheel disc hydraulic brake system. Technician A says seized caliper pistons may cause brake drag. Technician B says that rotors machined beyond the maximum diameter may cause brake pedal fade when you first apply the brakes. Who is right?
    A. A only
    B. B only
    C. Both A and B
    D. Neither A nor B (B1.9)

63. The diameter on a driveline parking brake drum is greater at the edges of the friction surface than in the center. Which of these faults describes this description?
    A. Bell-mouth drum
    B. Concave brake drum
    C. Drum out of round
    D. Convex drum (B1.11 and B2.5)

64. When a truck makes severe hard brake applications with hydraulic disc brakes that stop on light applications, excessive brake fade occurs. Technician A says this could be cause by low fluid level. Technician B says brake drums worn or machined beyond the maximum diameter could cause this condition. Who is right?
    A. A only
    B. B only
    C. Both A and B
    D. Neither A nor B (B2.1)

65. You are inspecting the brake pads on a truck equipped with hydraulic disc brakes. Technician A says you must remove the wheel in most cases to measure pad thickness. Technician B says if the brakes were applied shortly before measuring the lining to rotor clearance, this clearance will be greater than specified. Who is right?
   A. A only
   B. B only
   C. Both A and B
   D. Neither A nor B (B2.4)

66. A truck with a hydrovac brake booster requires excessive brake pedal effort. Technician A says the one-way check valve may be restricted in the vacuum hose to the hydrovac unit. Technician B says a sticking power piston in the hydrovac unit could be the cause. Who is right?
   A. A only
   B. B only
   C. Both A and B
   D. Neither A nor B (B3.1)

67. Cable-actuated parking brake systems are found in:
   A. most diagonally split brake systems.
   B. all air operated brake systems.
   C. all hydraulically operated systems.
   D. some air-over hydraulic parking systems. (B2.5)

6 Additional Test Questions for Practice

Additional Test Questions

Please note the letter and number in parentheses following each question. They match the overview in section 4 that discusses the relevant subject matter. You may want to refer to the overview using this cross-referencing key to help with questions posing problems for you.

1. What does the double check valve do?
   A. It ensures that air is applied in only one direction at one time.
   B. If one reservoir loses pressure, the double check valve alerts the functioning reservoir to supply air to both circuits.
   C. The double check valve alerts the driver as to which direction the air in a particular line is traveling at a given time.
   D. It ensures proper levels for governor cut-off pressures. (A1.12)

2. The majority of all brake applications are made at a brake chamber air pressure of:
   A. 100 psi or less.
   B. 80 psi or less.
   C. 40 psi or less.
   D. 20 psi or less. (A2.2)

3. Technician A says that the park and emergency braking system is a separate and distinct air circuit, completely isolated from the regular service air system. Technician B says that spring brake chambers use air pressure in the opposite way from the service brake chambers usage. Who is right?
   A. A only
   B. B only
   C. Both A and B
   D. Neither A nor B (A3.2)

4. You are servicing a truck's wheel bearings. Technician A says that you use a high quality high temperature grease when repacking is necessary. Technician B says it's okay to switch wheel bearings from one wheel to another. Who is right?
   A. A only
   B. B only
   C. Both A and B
   D. Neither A nor B (C2)

5. What is anticompounding?
   A. Reversing gladhand positions on the trailer and tractor lines
   B. Running the tractor in a bobtail position
   C. Rapidly releasing the air in the spring brake chamber to allow the brakes to be applied
   D. Simultaneous application of the service and emergency side of the spring brake chambers (A1.12)

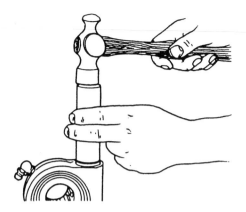

6. As shown in the figure above, you are installing a lip seal on a worm gear of a slack adjuster. Technician A says to install the lip of the seal towards the outside of the adjuster. Technician B says not to hit the seal after it reaches the bottom of the bore. Who is right?
   A. A only
   B. B only
   C. Both A and B
   D. Neither A nor B (A2.3)

7. Technician A says that power brake boosters can be found on hydraulically controlled brake systems. Technician B says that one finds power brake boosters on air-over-hydraulic systems. Who is right?
   A. A only
   B. B only
   C. Both A and B
   D. Neither A nor B (B3.2)

8. Technician A says that cam-actuated brake systems are the most popular brake system on heavy duty trucks today. Technician B says that cast brake shoes use an integral pin and roller design. Who is right?
   A. A only
   B. B only
   C. Both A and B
   D. Neither A nor B (A2.5)

9. The discharge valve in an air compressor:
   A. ensures proper direction of air into the wet tank.
   B. prevents the compressed air in the discharge line from returning to the cylinder bore as the intake and compression cycle is repeated.
   C. prevents the pressure in the cylinder from becoming too high, thus possibly damaging the internal parts of the compressor.
   D. allows compressed air to pass into the governor. (A1.6)

 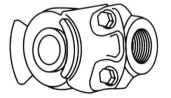

10. Shown in the figure above, what are gladhand seals made of?
    A. Rubber
    B. Polyurethane
    C. Plastic
    D. Rubber-covered aluminum (A1.21)

11. The pressure differential valve can also be known as the:
    A. pressure warning valve.
    B. system inproportional indicator.
    C. pressure valve.
    D. system effectiveness indicator. (B1.7)

12. An ABS system on a truck is being inspected. Technician A says that there is generally one modulator per axle, which means that each axle brakes as one unit. Technician B says that a truck's ABS brake system is designed to handle a "split coefficient" situation when that occurs. Who is right?
    A. A only
    B. B only
    C. Both A and B
    D. Neither A nor B (A4.3)

13. What is the most common design/type in use for brake shoes?
    A. Fabricated shoes
    B. Cast brake shoes
    C. Cam-actuated shoes
    D. Wedge-actuated shoes (B2.3)

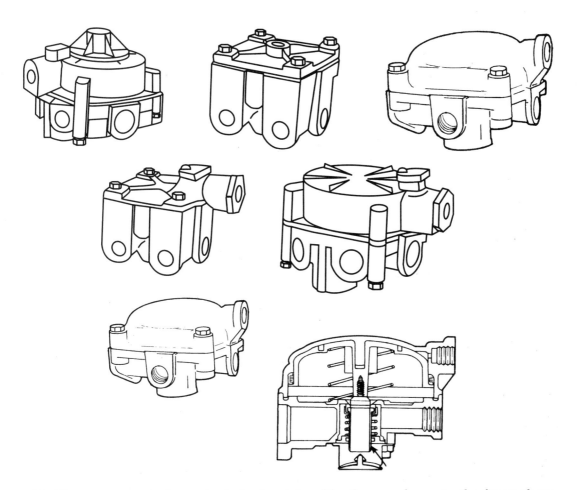

14. How many ports does a typical relay valve, like the one shown in the figure above, have for a tandem axle?
    A. Two
    B. Four
    C. Five
    D. Seven (A1.15)

15. You are testing a parking brake system on a truck equipped with air brakes. Technician A says that when the primary or secondary system fails, federal law requires that the parking brakes must release at LEAST once. Technician B says that in order for the truck to be mobile, air must release from the spring brake chambers. Who is right?
    A. A only
    B. B only
    C. Both A and B
    D. Neither A nor B   (A3.1)

16. In the S-cam foundation brake, the final link between the air system and the foundation brake is:
    A. the chamber pushrod.
    B. the slack adjuster.
    C. the camshaft.
    D. the chamber mounting plate.   (A2.4)

17. Technician A says that it is friction built up in or on the drum/rotor that actually stops a vehicle's motion. Technician B says it is the shoes/pads being applied against the drums/rotors that actually stop a vehicle's motion. Who is right?
    A. A only
    B. B only
    C. Both A and B
    D. Neither A nor B   (theory)

18. Technician A says that one of the factors affecting air timing is the size of the lines. Technician B says that if a trailer service system has an air fitting that is the wrong application (i.e., too small), then the air timing/balance will be affected. Who is right?
    A. A only
    B. B only
    C. Both A and B
    D. Neither A nor B   (A1.21)

19. A truck with a hydraulic braking system is being overhauled. Technician A uses a vacuum gauge to test master cylinder output. Technician B says that all the wheel cylinders must be replaced when an overhaul is being performed. Who is right?
    A. A only
    B. B only
    C. Both A and B
    D. Neither A nor B   (B1.4)

20. Technician A says that a slack adjuster must be adjusted so the angle between it and the chamber pushrod is 90° when the brakes are fully applied. Technician B says that the chamber pushrod should be adjusted with the shortest possible stroke, without dragging the brakes. Who is right?
    A. A only
    B. B only
    C. Both A and B
    D. Neither A nor B   (A2.2)

21. Which of these is the LEAST likely cause of wheel bearing failure?
    A. overloading
    B. contamination
    C. a damaged race/axle housing
    D. improper lubricant   (C2)

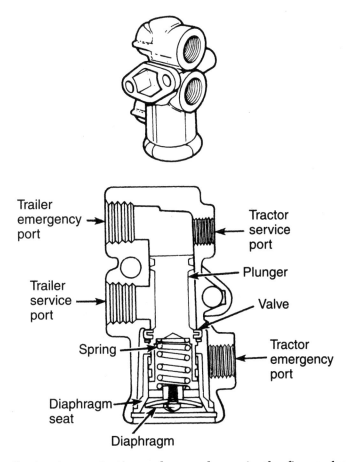

22. What does the tractor protection valve, as shown in the figure above, do?
    A. Ensures against air leakage at the gladhands
    B. Provides a means of preserving air pressure in sufficient amount to stop the tractor in event of a trailer breakaway
    C. Provides a means of holding off the spring brake in case of a spring brake chamber air loss
    D. It senses that the tractor is running without a trailer and automatically reduces the amount of air pressure that can be applied to the tractor's drive axle(s) (A1.18)

23. A power booster:
    A. can have dual diaphragm.
    B. is capable of converting hydraulic pressure into mechanical pressure.
    C. is nonrepairable.
    D. is exchangeable with other booster(s). (B3.2)

24. What should be used when flushing a hydraulic brake system?
    A. Water
    B. Mineral spirits
    C. Gasoline
    D. Alcohol (B10)

25. A heavy truck brake system is being inspected. Technician A says that the proportioning valve should be inspected every time the brakes are serviced. Technician B uses gauges ahead of and behind the proportioning valve to test its function. Who is right?
    A. A only
    B. B only
    C. Both A and B
    D. Neither A nor B (B1.6)

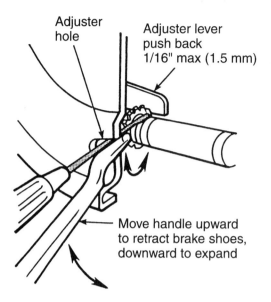

26. When adjusting the hydraulic drum brakes, as in the figure shown above, the self-adjusting mechanism should keep the lining-to-drum clearance to:
    A. 0.010–0.025 inch.
    B. 0.008–0.016 inch.
    C. 0.005–0.010 inch.
    D. 0.004–0.012 inch.
    (B2.3)

27. Cable-actuated parking brake systems are found in:
    A. most diagonally split brake systems.
    B. all air operated brake systems.
    C. all hydraulically operated systems.
    D. some air-over-hydraulic parking systems.
    (B2.5)

28. A hydromax power brake booster is being serviced. Technician A says that if normal flow from the hydraulic brake pump is interrupted, the electric motor pump provides power for reserve stops. Technician B says manual braking is possible if the power and reserve systems fail, but the braking will be extremely difficult and only emergency use of the vehicle is recommended. Who is right?
    A. A only
    B. B only
    C. Both A and B
    D. Neither A nor B
    (B3.1)

29. What is electronic braking?
    A. The addition of an electronic computer to an ABS braking system.
    B. An integrated braking system ensuring that the correct brake pressure is distributed to each wheel at any given moment.
    C. A synchronous ABS-based brake system allowing fully graduated control and performance through a series of digital relays.
    D. A braking system that uses an electronic proportional signal traveling by wire from the brake pedal to a solenoid cluster pack, then on to a brake computer.
    (A4.2)

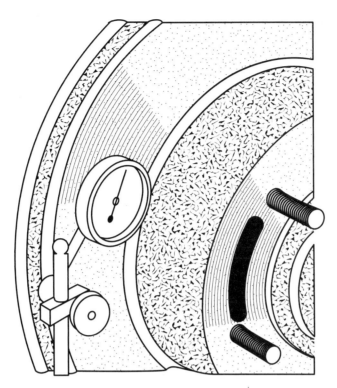

30. An air disc brake system is being inspected. Technician A says that thickness variation is being checked in the figure above. Technician B says the caliper pistons are responsible for stopping the vehicle by forcing the pads against the rotors. Who is right?
    A. A only
    B. B only
    C. Both A and B
    D. Neither A nor B   (A2.8)

31. Why is it very difficult to have the same actuation pressure on all axles of a combination vehicle?
    A. Because there are more valves between the foot valve and the trailer axles than the tractor axles, thus increasing restrictions and lowering those respective pressures.
    B. Because the increased length of the lines to the trailer causes a corresponding pressure drop or decrease.
    C. The same actuation pressure is impossible to achieve because both the tractor and trailer air systems are two separate and distinct systems.
    D. Because the treadle valve operates the tractor air system, while the trailer supply valve operates the trailer air system.   (A1.8)

32. What is the LEAST likely operation (secondary role) of an air compressor?
    A. Provides air to windshield wipers
    B. Assists steering units by compressed air
    C. Drives a fuel pump
    D. Provides air for heater   (A1.6)

33. Why should a slack adjuster need adjusting?
    A. Wedge-type brakes by design require frequent adjusting; otherwise rapid uneven lining wear will ensue.
    B. The proper free stroke is the reason for adjusting drum to lining clearance on cam-type slack-adjusting brake systems.
    C. If pushrod length is too short, the brakes will drag; thus the clevis to pushrod adjustment is critical.
    D. The angle between the brake chamber pushrod and the centerline of the slack adjuster must be 90° with the brakes fully applied. (A2.3)

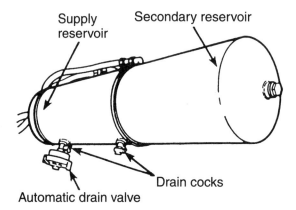

34. As shown in the figure above; Technician A says a wet tank is called this name because moisture forms in this tank as the compressed air cools. Technician B says that in a two-reservoir system (one supply and one service), the two reservoirs work independently of each other; thus, if pressure in one reservoir fails, the second reservoir remains operable. Who is right?
    A. A only
    B. B only
    C. Both A and B
    D. Neither A nor B (A1.3)

35. What would happen to a wheel bearing assembly if the wheel bearing adjusting nut were not backed off after initial torquing?
    A. Nothing as long as the bearing receives the proper lubrication.
    B. Nothing, as long as the initial procedure of torquing and correct torque values were followed according to the manufacturer's specifications.
    C. The bearing will seat itself after an initial "run-in" period of about 100 miles.
    D. The bearing will run hot and be damaged, possibly causing the wheel to lock or come off during operation. (C3)

36. When replacing the pads on both brakes of a single axle or all four brakes of a tandem axle you should:
    A. replace all at the same time.
    B. replace only one side at a time.
    C. replace only the worn pads, while measuring the rest.
    D. resurface all discs that have had their pad(s) replaced. (A2.8)

37. A truck is brought in with combination lining sets. Technician A says that combination linings are linings on the same drum that are of different sizes. Technician B says differences in construction (i.e., riveted, bolted) is a definition of combination linings. Who is right?
    A. A only
    B. B only
    C. Both A and B
    D. Neither A nor B (A2.8)

38. Never use brake fluid out of a container:
    A. that is made out of glass.
    B. that has been tightly stored.
    C. that has been used to store any other liquid.
    D. that has had brake fluid in it for more than one year. (B1.10)

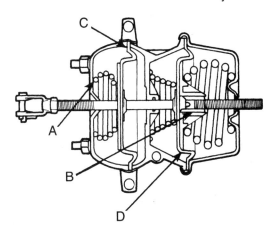

39. What is the part marked "B" in the figure shown above?
    A. S-cam
    B. Power spring
    C. Check valve
    D. Pressure plate (A2.2)

40. A wheel sensor from an ABS system is being inspected. Technician A says that as the toothed wheel rotates, the sensor generates a simple DC signal. Technician B says that the waveform signal generated is sinusoidal and is proportional to wheel speed. Who is right?
    A. A only
    B. B only
    C. Both A and B
    D. Neither A nor B (A4.5)

41. What is one of the most important aspects of air compressor preventative maintenance?
    A. The induction of clean air
    B. Maintain proper oil level
    C. Ensure secure mounting of unit
    D. Maintain efficient governor operation (A1.6)

42. Technician A says that a hydraulic disc brake caliper can have two pistons. Technician B says that a disc brake caliper can have four pistons. Who is right?
    A. A only
    B. B only
    C. Both A and B
    D. Neither A nor B (B2.4)

43. A return line is being installed to a hydraulic power booster. Technician A says the line must be a hydraulic hose conforming to SAE J189. Technician B uses a pressure bleeder to bleed the hydro-boost system. Who is right?
    A. A only
    B. B only
    C. Both A and B
    D. Neither A nor B (B3.1)

44. What is the generally acceptable pressure differential between the tractor rearmost axle and the trailer rearmost axle?
    A. 2 psi
    B. 4 psi
    C. 6 psi
    D. 8 psi (A1.2)

45. Some manufacturers use two different free stroke settings for their air disc brakes. Which if the following is one of the settings?
    A. Full-clevis free stroke
    B. Half-clevis free stroke
    C. Full free stroke
    D. initial free stroke (A2.7)

46. Technician A says both front rear wheel bearings are adjustable. Technician B says you use a brass drift punch to remove a rear wheel bearing. Who is right?
    A. A only
    B. B only
    C. Both A and B
    D. Neither A nor B (C2)

47. Hydraulic single piston disc brakes are said to have:
    A. servo action.
    B. nonservo action.
    C. energizing action.
    D. action and reaction. (A2.5)

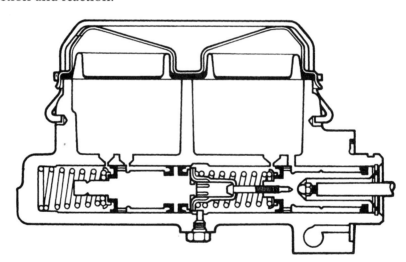

Cross-section of a typical dual master cylinder

48. As shown in the figure above, after an application, what component could prevent the master cylinder piston from returning the brake pedal to its original unapplied position?
    A. A check valve
    B. The metering valve
    C. A sliding rubber seal
    D. A return spring (B1.4 & B1.3)

49. An air brake system is being inspected. Technician A says the pushrod spring in combination with the brake shoe return spring releases the brakes. Technician B says the clevis threads onto the slack adjuster arm. Who is right?
    A. A only
    B. B only
    C. Both A and B
    D. Neither A nor B (A2.2)

Additional Test Questions for Practice Additional Test Questions 79

50. Technician A says that on brakes with vented discs, the inboard pad could be thicker than the outboard pad. Technician B says that on brakes with solid discs, the inboard and outboard pads could be the same thickness. Who is right?
    A. A only
    B. B only
    C. Both A and B
    D. Neither A nor B (B2.4)

51. A tractor has a problem with its air pressure rising above normal. Technician A says the check valve on the governor could be stuck closed. Technician B says there could be too much clearance at the compressor unloading valves. Who is right?
    A. A only
    B. B only
    C. Both A and B
    D. Neither A nor B (A1.7)

52. Technician A says that a dual power hydraulic brake booster could have one diaphragm. Technician B says that a dual power hydraulic brake booster could have two diaphragms. Who is right?
    A. A only
    B. B only
    C. Both A and B
    D. Neither A nor B (B3.2)

53. A wheel bearing with a unitized seal is being replaced. Technician A says any time a wheel with a unitized seal is removed, the seal *must* be replaced. Technician B says that wheel seals should be removed using a brass drift and a hammer. Who is right?
    A. A only
    B. B only
    C. Both A and B
    D. Neither A nor B (C2)

54. What is the minimum performance (psi/sec) a compressor should have?
    A. 85–100 psi in 25 seconds or less
    B. 60–80 psi in 20 seconds or less
    C. 90–110 psi in 30 seconds or less
    D. 50–110 psi in 25 seconds or less (A1.6)

55. Which of the following is the LEAST likely cause of uneven braking?
    A. Improper axle mounting
    B. A stuck relay valve
    C. A broken brake shoe return spring
    D. Brake chamber diaphragm failure (A1.1)

56. Technician A says that when performing a brake overhaul on a hydraulic rear drum brake system, you replace the return springs. Technician B says the maximum allowable out-of-round specification on a brake drum should be 0.025 inch. Who is right?
    A. A only
    B. B only
    C. Both A and B
    D. Neither A nor B (B2.2)

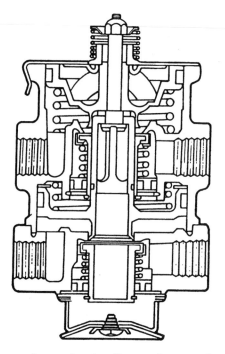

57. The application valve, as shown in the figure above, is being repaired. Technician A says that the brake valve has two pistons: a reaction or rear modulating piston, and a primary or front modulating piston. Technician B says that a reaction spring between the two pistons of a brake valve is responsible for the basic operation of the valve. Who is right?
    A. A only
    B. B only
    C. Both A and B
    D. Neither A nor B   (A1.11)

58. Technician A says that the maximum recommended discharge air temperature is 300° F. Technician B says that most air compressors are liquid cooled.
    Who is right?
    A. A only
    B. B only
    C. Both A and B
    D. Neither A nor B   (A1.6)

59. A full-floating wheel assembly is being inspected. Technician A says that on disc wheels, the wheel and hub are combined into a single unit. Technician B says a wheel bearing may be locked in place by either an adjusting and lock nut assembly, split forged nut, or castellated nut and cotter pin assembly.
    Who is right?
    A. A only
    B. B only
    C. Both A and B
    D. Neither A nor B   (C1)

60. During air brake diagnosis the application valve is suspected to be the cause of slow apply. When diagnosing the brake application valve, Technician A says the air pressure at the brake application valve delivery port is proportional to treadle movement. Technician B says with the brakes applied, a 1 inch (2.54 cm) bubble in 3 seconds at the brake application valve exhaust port indicates excessive leakage. Who is right?
    A. A only
    B. B only
    C. Both A and B
    D. Neither A nor B   (A1.11)

61. Technician A says master cylinder parts may be washed in an oil-based solvent. Technician B says an aluminum master cylinder bore may be honed if it is scored. Who is correct?
    A. A only
    B. B only
    C. Both A and B
    D. Neither A nor B   (B1.4)

62. When discussing quick-release valve diagnosis, technician A says that with the service brakes applied, acceptable leakage at the quick-release valve exhaust port is a 3 inch (2.54 cm) bubble in 1 second. Technician B says that when the service brakes are applied and released, the air pressure in the front brake chambers should be released immediately. Who is right?
    A. A only
    B. B only
    C. Both A and B
    D. Neither A nor B   (A1.1 & A1.16)

63. During a brake inspection, a technician tests the air supply system and finds that the buildup is slow. Which of these is the most likely cause?
    A. A clogged compressor inlet filter
    B. A leak in a brake chamber
    C. An air leak in the cab suspension
    D. A restriction in the governor line   (A1.2)

64. An automatic reservoir drain valve has which of these attributes?
    A. It opens when the reservoir pressure reaches the governor pressure and is unloaded.
    B. It is mounted in a threaded fitting near the top of the reservoir.
    C. It opens by decreasing supply reservoir pressure.
    D. It drains water if the sump pressure is 2 psi higher than reservoir pressure.   (A1.3)

65. When you are adjusting the tension of the compressor belt, and you notice contact at the bottom of the pulley with the belt, you should:
    A. decrease tension to original specification level and recheck contact surface for excessive wear.
    B. measure the level of deflection at a section of belt that is longest between pulleys.
    C. replace the belt.
    D. ignore it because some applications require this additional contacting area.   (A1.4)

66. You are replacing an air compressor drive on a truck with a two-stroke diesel engine. Technician A says it is necessary to retime the engine when this replacement process is complete. Technician B says you inspect the gear for worn or chipped teeth. Who is right?
   A. A only
   B. B only
   C. Both A and B
   D. Neither A nor B (A1.4)

67. The dual wheel assembly of a tractor axle is being reinstalled. Technician A says you put a light coating lubricant on the spindle and place a seal protector over the threads. Technician B says rear wheel hub assemblies do not use oil filler plugs to fill the wheel cavities. Who is right?
   A. A only
   B. B only
   C. Both A and B
   D. Neither A nor B (C1)

68. When testing and adjusting the governor cut-out and cut-in pressures, which of these statements apply?
   A. The cut-out pressure may be increased by rotating the adjusting screw counterclockwise.
   B. You turn the adjusting screw clockwise to increase cut-in pressure.
   C. You turn the adjustment screw ¼ turn to change cut-in or cut-out pressure 7 psi (48.27 Kpa).
   D. If you restrict the airline from the supply reservoir to the governor, the cut-out pressure is decreased. (A1.7)

69. Technician A says that daily draining of air reservoirs that are not equipped with automatic drain valves is highly recommended. Technician B says that you should check all automatic drain valves and moisture removing devices periodically for proper operation. Who is right?
   A. A only
   B. B only
   C. Both A and B
   D. Neither A nor B (A1.9)

70. What is the LEAST likely cause of water in air reservoirs?
   A. Dented or kinked air lines
   B. Lack of periodic drainage of supply tanks
   C. Damaged air reservoir tanks
   D. Continuous operation in very humid conditions (A1.10)

71. When testing the operation of the trailer control valve, a technician should:
   A. move the handle to the fully applied position and record the air pressure.
   B. drain the air system to 0 psi, and then allow pressure to build up again.
   C. record the gauge reading of the air pressure first, and then drain the system.
   D. listen for air leakage around the handle as the handle is moved to the applied position. (A1.14)

72. A truck is experiencing improper trailer service brake operation. Technician A says that a defective trailer control valve could be the cause. Technician B says that excessive compressor cycling is a symptom of this same condition. Who is right?
   A. A only
   B. B only
   C. Both A and B
   D. Neither A nor B (A1.14)

73. When diagnosing a service brake relay valve, all of the following apply **EXCEPT:**
    A. inlet valve leakage is tested with the service brakes released.
    B. apply a soap solution to the area around the inlet and exhaust valve retaining ring to check exhaust valve leakage.
    C. exhaust valve leakage is tested with the brakes applied.
    D. the control port on the service brake relay valve is connected to the supply port on the brake application valve. (A1.15)

74. When considering a tractor service brake relay valve:
    A. air lines are connected from the service brake relay valve delivery ports to the rear axle service brake chambers.
    B. when the brakes are released, the inlet valve is open in the service brake relay valve.
    C. when the brakes are released, the exhaust valve is closed in the service brake relay valve.
    D. the service brake relay valve is in a balanced position when the air pressure in the rear brake chambers equals reservoir pressure. (A1.15)

75. All of these statements about limiting quick-release valve diagnosis are true **EXCEPT:**
    A. with the control valve in the dry road position, the limiting quick-release valve reduces air pressure to the front brakes.
    B. with the control valve in the slippery road position and the service brakes applied, leakage at the control valve exhaust port should not exceed a 1-inch (2.54 cm) bubble in 3 seconds.
    C. with the control valve in the slippery road position and the service brakes applied, leakage at the limiting quick-release valve exhaust port should not exceed a 1-inch (2.54 cm) bubble in 3 seconds.
    D. with the control valve in the slippery road position and the service brakes applied, it limits front brake pressure to 50 percent of the application valve pressure. (A1.17)

76. Technicians are discussing quick-release valves. Technician A says that a bad tractor protection valve will cause wheel lockup during a brake application. Technician B says a defective quick-release valve can cause slow front pressure buildup and poor stopping ability. Who is right?
    A. A only
    B. B only
    C. Both A and B
    D. Neither A nor B (A1.16)

77. With 50 psi (344.75 Kpa) supplied from the brake application valve to the limiting valve inlet port, the pressure read on a test gauge at the limiting valve outlet port should be:
    A. 40 psi (275.8 Kpa).
    B. 15 psi (103.42 Kpa).
    C. 25 psi (172.37 Kpa).
    D. 30 psi (206.85 Kpa). (A1.17)

78. Technicians are diagnosing the tractor spring brake control valve. Technician A says if the rear axle reservoir is drained to zero air pressure and the brake application valve is applied, the tractor air brake pressure should be at 120 psi. Technician B says when the tractor parking brakes are applied, the air pressure at the tractor spring brake valve delivery port should drop quickly to zero.
    Who is right?
    A. A only
    B. B only
    C. Both A and B
    D. Neither A nor B (A1.19)

79. While discussing single and double check valves, technician A says single check valves are connected between the primary and secondary reservoirs. Technician B says a double check valve allows air pressure to flow from the lowest of two pressure sources. Who is correct?
    A. A only
    B. B only
    C. Both A and B
    D. Neither A nor B   (A1.9)

80. The red low-pressure warning light on a vehicle does not go out after startup. Technician A says that the problem may be caused by a loss of one section of the dual system. Technician B says that the problem may be caused by a loss of the complete air brake system. Who is right?
    A. A only
    B. B only
    C. Both A and B
    D. Neither A nor B   (A1.20)

81. Technician says that an inoperative warning light and buzzer below 60 psi is caused by a defective bulb. Technician B says that an inaccurate dash pressure gauge can be the cause of a warning light on above 60-psi supply pressure. Who is right?
    A. A only
    B. B only
    C. Both A and B
    D. Neither A nor B   (A1.20)

82. All of the following can be causes of torque imbalance **EXCEPT**:
    A. oil-soaked linings.
    B. nonuniform adjustment.
    C. nonuniform chamber sizes.
    D. an overloaded vehicle.   (A2.1)

83. The brake linings are worn and there is excessive clearance between the linings and the brake drum on a cam-type foundation brake. Technician A says this condition may cause the cam to roll over during a brake application. Technician B says this condition may cause loss of steering control. Who is right?
    A. A only
    B. B only
    C. Both A and B
    D. Neither A nor B   (A2.1)

84. When assembling a camshaft all of the following apply **EXCEPT**:
    A. both camshaft seal lips must face toward the slack adjuster.
    B. use the proper drive when installing camshaft seals.
    C. lubricate the cam head with petroleum jelly.
    D. measure camshaft radial movement.   (A2.4)

85. While discussing wedge brake diagnosis and service, technician A says installing a new wedge with a different angle does not affect brake operation. Technician B says no lubricant should be used on the wedge or wedge ramps. Who is correct?
    A. A only
    B. B only
    C. Both A and B
    D. Neither A nor B   (A2.6)

86. When servicing wedge brakes all of these apply **EXCEPT:**
   A. minor scoring may be removed from the actuator bores with fine emery cloth.
   B. if the threaded actuator opening is not chamfered, install the collet nut so the flat side faces away from the actuator threaded opening.
   C. the clearance between each wedge brake shoe lining and the drum should be 0.020 to 0.040 inch (0.51 to 1.02 mm).
   D. if the threaded actuator opening is chamfered, the taper on this collet nut faces the taper on the actuator opening. (A2.6)

87. All of the following are parts of an air disc brake **EXCEPT:**
   A. hydraulic hoses.
   B. torque plates.
   C. calipers.
   D. rotors. (A2.7)

88. New brake linings are being installed on a truck. Technician A says you use the same procedure with bolted linings as with riveted ones. Technician B always greases the platforms on the backing plate before installing new shoes/linings. Who is right?
   A. A only
   B. B only
   C. Both A and B
   D. Neither A nor B (A2.8)

89. Technician A says that severe heat cracking will cause rapid lining wear. Technician B says that brake drums (over 14 inches in diameter), may be machined 0.130 inch over normal diameter safely. Who is right?
   A. A only
   B. B only
   C. Both A and B
   D. Neither A nor B (A2.9)

90. What can extreme rotor thickness variation result in?
   A. Low brake output
   B. Pedal pulsation
   C. Loss of brake fluid
   D. Loss of directional control (A2.9)

91. Which of the following is the LEAST likely cause of a parking brake that will not hold?
   A. Slack adjuster alignment/adjustment
   B. A defective self-adjusting star wheel defective
   C. Foundation brakes that need repair/replacing
   D. Weak spring brake chamber springs (A3.1)

92. The control line is more commonly called the:
   A. apply line.
   B. emergency line.
   C. park line.
   D. trailer line. (A3.2)

93. While discussing a tractor park valve, technician A says the tractor park valve has a red, diamond-shaped knob that is pulled out to apply the trailer parking brakes. Technician B says the tractor park valve may be used to apply the trailer parking brakes before uncoupling the trailer from the tractor. Who is right?
   A. A only
   B. B only
   C. Both A and B
   D. Neither A nor B (A3.3)

94. Technician A says that the trailer air supply valve allows air to enter the trailer brake system when its knob is pulled outward. Technician B says that the knob of the park control (system park) valve must be pulled out before the trailer air supply valve can receive air at the control port. Who is right?
    A. A only
    B. B only
    C. Both A and B
    D. Neither A nor B (A3.3)

95. Technician A says that pressure below 45 psi in the air brake system will cause the knob of the trailer air supply valve to pop out. Technician B says that pressure below 45 psi in the trailer air brake system will cause the control valve inside the tractor protection valve to close. Who is correct?
    A. A only
    B. B only
    C. Both A and B
    D. Neither A nor B (A3.4)

96. When the trailer air supply valve knob is pulled outward and the park control (system park) valve knob is pushed inward:
    A. the trailer air supply line is charged.
    B. the tractor parking brakes are applied.
    C. air is exhausted through the park control valve.
    D. the trailer parking brakes are applied. (A3.4)

97. All of the following could cause failure of the ABS system **EXCEPT:**
    A. shorted solenoid coils.
    B. leaking hydraulic lines.
    C. blocked exhaust solenoid ports.
    D. an open circuit in the ECU. (A4.1)

98. The National Transportation Safety Board requires:
    A. that all trucks with a gross vehicle weight rating (GVWR) over 10,000 lbs. have dual-action braking.
    B. that all new trucks, trailers, and heavy tractors built after 1996 be equipped with an antilock braking system.
    C. that all trucks over 20,000 pounds GVWR have type 35 Rotochambers.
    D. that all trucks, buses, heavy tractors, and trailers be equipped with forged steel brake chamber pushrods that have at least 50,000-pound tensile strength. (A4.1)

99. Which of the following is the LEAST likely used ABS system on heavy duty trucks?
    A. Eight-channel system
    B. Six-channel system
    C. Four-channel system
    D. Two-channel system (A4.4)

100. Which of the following conditions can cause brake pedal fade?
    A. a seized brake caliper piston
    B. brake drum machined beyond its limit
    C. leakage past the master cylinder cups
    D. air in the hydraulic system (B1.1)

101. What should the inside of a new hydraulic brake line be cleaned with?
    A. Soap and water
    B. Isopropyl alcohol
    C. Mineral spirits
    D. Hydraulic brake fluid (B1.5)

102. On a truck equipped with disc/drum brakes, the front brakes grab quickly when light pedal pressure is applied. This could be caused by a bad:
   A. proportioning valve.
   B. pressure differential valve.
   C. Metering valve.
   D. residual check valve. (B1.6 & B1.7)

103. A hydraulic brake system is being bled with a metering valve. Technician A applies a device to hold the metering valve open when he uses a pressure bleeder to bleed the system. Technician B bleeds the brakes manually with no problems. Who is right?
   A. A only
   B. B only
   C. Both A and B
   D. Neither A nor B (B1.10)

104. The diameter on a driveline parking brake drum is greater at the edges of the friction surface than in the center. Which of these faults describes this description?
   A. bell-mouth drum
   B. concave brake drum
   C. drum out of round
   D. convex drum (B1.11 & B2.5)

105. Technician A says that vacuum-operated power boosters are used on heavy duty trucks. Technician B says that hydraulic boosters are used on heavy duty trucks. Who is right?
   A. A only
   B. B only
   C. Both A and B
   D. Neither A nor B (B3.2)

106. What action would occur if the reserve electrical motor for the hydraulic power brake system failed?
   A. You will have no reserve stops without the engine running.
   B. There will be no reserve electrical power in case of an electrical failure in the booster.
   C. Fluid circulation in the steering gear will fail.
   D. Braking power will be reduced with the engine running. (B3.3)

107. Technician A says that when working on air-over-hydraulic brake systems, an antifreeze valve may be used. Technician B says that a limiting valve is often used on rear axle brakes. Who is right?
   A. A only
   B. B only
   C. Both A and B
   D. Nether A nor B (B3.3)

108. A heavy truck's wheel bearings:
   A. should be repacked every 10,000 miles.
   B. are of a semifloating design.
   C. should be inspected at every PM.
   D. are sometimes constructed of aluminum. (C2)

109. When adjusting wheel bearings, Technician A says torque the adjusting nut to 50 ft.-lbs., then back off the nut one-sixth to one-third turn and install the lock ring. Technician B says backing off the nut will cause the bearing to overheat. Who is right?
   A. A only
   B. B only
   C. Both A and B
   D. Neither A nor B (C3)

110. Which of the following causes are correct when diagnosing hydraulic brake systems?
    A. A chassis vibration during braking is caused by excessive radial tire runout.
    B. Brake grab on one wheel is caused by an improper brake pedal pushrod adjustment.
    C. Brake drag is caused by glazed brake linings.
    D. A low spongy pedal is caused by rotors that are machined too thin. (B1.1)

111. The drag on a truck's drum brakes increases after each application until they lockup. Bleeding the system restores normal operation for a short time; then the drag increases until the brakes lockup again. Which of these could be the cause?
    A. a swollen master cylinder cup
    B. an open compensation port
    C. the wrong brake shoes
    D. a weak brake shoe return spring (B1.2)

112. Which of the following procedures is correct when pressure bleeding a hydraulic brake system?
    A. You bleed the right front caliper or wheel cylinder first.
    B. The pressure bleeder is pressurized to 20 to 25 psi (137.9–172.37 Kpa).
    C. If equipped, the metering valve must be closed.
    D. You bleed the left rear caliper or wheel cylinder last. (B1.2)

113. If the floating caliper on a hydraulic brake truck does not slide freely, which of these conditions will result?
    A. excessive brake force
    B. excessive brake pad wear
    C. reduced braking force
    D. the brakes will grab (B1.9)

114. When servicing disc brake assemblies, which of these applies to the pads?
    A. Thickness is indicated by a code on the lining edge.
    B. All semimetallic linings are edge-stamped "FF."
    C. All asbestos-type linings are edge-stamped "EE."
    D. Linings should use the same edge brand as OEM. (B1.9)

115. Technician A says it's OK to machine oversize drums as long as the drum's manufacturer's recommendation for machining dimensions are followed. Technician B says that turning an oversize drum will sacrifice its strength. Who is right?
    A. A only
    B. B only
    C. Both A and B
    D. Neither A nor B (B2.2)

116. When testing brake application valve leakage, Technician A says that the brake application pressure should be 80 psi (553.6 Kpa). Technician B says that this test is done every 50,000 miles. Who is correct?
    A. A only
    B. B only
    C. Both A and B
    D. Neither A nor B (A1.11)

117. All of the following steps in removing an application valve are correct **EXCEPT**:
    A. the truck should be on a level surface.
    B. mark or label the brake lines.
    C. mark the valve body in relation to the mounting plate.
    D. maintain brake system pressure. (A1.11)

118. When testing single and double check valves:
    A. With air pressure applied to the inlet side of a single check valve, leakage must not exceed specifications at the outlet side.
    B. With air pressure applied to one of the inlet ports on a double check valve, the air pressure should increase slowly at the outlet port.
    C. With air pressure applied to one inlet port in a double check valve, the test gauge at the opposite inlet port should indicate air pressure.
    D. When air pressure is released at one inlet port on an double check valve, the air pressure should decrease slowly at the outlet port. (A1.12)

119. When diagnosing a pressure protection valve, technician A says that when the air pressure is reduced to zero on the delivery side of the pressure protection valve, the air pressure should decrease to zero on the supply side of this valve. Technician B says that a 2-inch bubble in 5 seconds is acceptable around the cap on the pressure protection valve. Who is right?
    A. A only
    B. B only
    C. Both A and B
    D. Neither A nor B (A1.20)

120. While discussing tractor protection valves, technician A says when the trailer air supply valve is pulled outward, the air pressure to the tractor protection valve is open. Technician B says if a trailer breakaway occurs, service air pressure from the application valve can no longer pass through the tractor protection valve. Who is right?
    A. A only
    B. B only
    C. Both A and B
    D. Neither A nor B (A1.18)

Appendices

Answers to the Test Questions for the Sample Test Section 5

1. A	18. B	35. C	52. B
2. B	19. A	36. A	53. D
3. C	20. D	37. D	54. C
4. B	21. C	38. C	55. C
5. D	22. B	39. B	56. B
6. A	23. A	40. A	57. C
7. C	24. D	41. C	58. D
8. D	25. C	42. A	59. C
9. C	26. C	43. D	60. A
10. B	27. B	44. B	61. A
11. D	28. A	45. A	62. A
12. B	29. D	46. A	63. D
13. C	30. B	47. B	64. B
14. C	31. C	48. D	65. A
15. D	32. C	49. C	66. C
16. D	33. B	50. A	67. D
17. B	34. D	51. C	

Explanations to the Answers for the Sample Test Section 5

Question #1
Answer A is correct because the spring brake chambers should be "caged" before disconnecting any air line or hose due to safety reasons.
Answer B is wrong because the spring brakes apply when the park valves cause the relay valves to drain all air pressure.
Answer C is wrong because only one technician is right.
Answer D is wrong because one technician is right.

Question #2
Answer A is wrong because a slack adjuster is considered a part of the foundation brakes.
Answer B is correct because the component shown is a combination valve.
Answer C is wrong because the treadle valve is air-operated, and is mounted on the floor of the cab.
Answer D is wrong because the tractor protection valve is an air-operated and controlled valve.

Question #3
Answer A is a good choice because the ABS ECU uses a series of LEDs in the diagnostic window to indicate fault codes in the ECU. Yet, it is wrong because both technicians are right.
Answer B is also a good choice because the number of LEDs in the diagnostic window varies depending on the system. Yet, it is wrong because both technicians are right.
Answer C is correct because of the above reasons and because both technicians are right.
Answer D is wrong because both technicians are right.

Question #4
Answer A is wrong because slack adjusters are part of a truck's foundation brakes.
Answer B is correct because hydraulic or pneumatic lines are part of the control system.
Answer C is wrong because brake pads are part of a truck's foundation brakes.
Answer D is wrong because brake chambers are part of a truck's foundation brakes.

Question #5
Answer A is wrong because the quick-release valve speeds up the exhaust air from the air chambers.
Answer B is wrong because the relay valve speeds up the application and release of the brakes.
Answer C is wrong because the trailer supply valve works with the parking control valve to provide tractor protection and control trailer parking brake application.
Answer D is correct because the bobtail proportioning valve automatically reduces the amount of air pressure that can be applied to the tractor drive axle(s).

Question #6
Answer A is correct because the clevis should be correctly positioned on the pushrod in order to facilitate proper brake adjustment.
Answer B is wrong because the clevis is not keyed, but splined, onto the brake pushrod.
Answer C is wrong because only one technician is right.
Answer D is wrong because one technician is right.

Question #7
Answer A is a good choice because it is a good practice to change the brake fluid whenever you perform a major brake repair. Yet, it is wrong because both technicians are right.
Answer B is a good choice because you should immediately replace contaminated brake fluid. Yet, It is wrong because both technicians are right.
Answer C is correct because of the above reasons and because both technicians are right.
Answer D is wrong because both technicians are right.

Question #8
Answer A is wrong because piston rings are part of a brake system's air compressor.
Answer B is wrong because the crankshaft is a fundamental part of an air compressor.
Answer C is wrong because the discharge valve is part of an air compressor.
Answer D is correct because air compressors do not use roller bearings.

Question #9
Answer A is wrong because a condition causing a lowering of air pressure will subsequently reduce service brake application force.
Answer B is wrong because an incorrect or lack of adjustment to a slack adjuster will cause a reduction in overall brake application forces.
Answer C is correct because grease on shoe linings/faces will cause grabbing. It is the least likely cause of insufficient braking.
Answer D is wrong because a ruptured diaphragm will severely decrease an air brake system's service application force.

Question #10
Answer A is wrong because a wheel cylinder need not be replaced as often as brake shoes/linings.
Answer B is correct because drums and brake shoes are high-wear items.
Answer C is wrong because anchor pins generally do not wear out as fast brake shoes.
Answer D is wrong because return springs often last far longer than brake shoes/linings.

Question #11
Answer A is wrong because the ultimate pushrod travel will be decreased if the brake chamber pushrod is less than 90° from the slack adjuster arm.
Answer B is wrong because a brake chamber pushrod cannot be parallel to the slack adjuster to be functional.
Answer C is wrong because overall braking force will be decreased if there is more than a 90° angle between the brake chamber pushrod and the slack adjuster arm.
Answer D is correct the braking force is at the maximum if there is less than a 90° angle between the brake chamber pushrod and the slack adjuster arm.

Question #12
Answer A is wrong because a tractor/trailer combination is capable of more than three reservoirs.
Answer B is correct because you have three reservoirs (one supply and two service) on the tractor and the trailer for a total of six.
Answer C is wrong because nine reservoirs are not needed for a typical combination vehicle.
Answer D is wrong because twelve reservoirs are not needed in a combination vehicle due to design and space parameters.

Question #13
Answer A is a good choice because brinelling appears as indentations across the bearing raceway. Yet, it is wrong because both technicians are right.
Answer B is also a good choice because brinelling occurs when the bearing is not rotating. Yet, it is wrong because both technicians are right.
Answer C is correct because of the above reasons and because both technicians are right.
Answer D is wrong because both technicians are right.

Question #14
Answer A is wrong because poor alignment of the foundation brakes linkage will effectively decrease braking power produced.
Answer B is wrong because a broken or defective chamber return spring will increase brake release rate.
Answer C is correct because broken or weak brake return springs will increase brake release rate.
Answer D is wrong. An obstruction in a brake chamber will decrease overall pushrod travel, thereby decreasing the rate of ultimate braking force.

Question #15
Answer A is wrong because air must be built up in the service chamber to apply the service brake.
Answer B is wrong because when air is released from the spring brake chamber, the spring brakes apply.
Answer C is wrong because a relay valve redirects air to the spring brake chambers.
Answer D is correct because when the valve is in the position shown, it releases the trailer parking brakes with supply air pressure above 85 psi.

Question #16
Answer A is wrong because pressure differential valves automatically re-center themselves upon application of the brakes after repairs have been made.
Answer B is wrong because you do not replace the proportioning valve after resetting the pressure differential valve.
Answer C is wrong because neither technician is right.
Answer D is correct because both technicians are wrong.

Question #17
Answer A is wrong because a worm gear is part of a slack adjuster assembly.
Answer B is correct because there are no pushrods in the slack adjuster assembly. The pushrod is part of the brake chamber.
Answer C is wrong because slack adjusters contain actuator pistons as part of the actuator assembly.
Answer D is wrong because the clevis is part of the slack adjuster assembly.

Question #18
Answer A is wrong because the fail relay coil is installed between a vehicle power source and the modulator assembly.
Answer B is correct because power for the fail light flows through the normally closed relay contacts.
Answer C is wrong because only one technician is right.
Answer D is wrong because one technician is right.

Question #19
Answer A is correct because overtorqued is not a common brake drum wear condition.
Answer B is wrong because a concave drum can be a possible brake drum wear condition.
Answer C is wrong because scoring of a drum is a commonly found wear condition.
Answer D is wrong because a "threaded" drum is a commonly found wear condition.

Question #20
Answer A is wrong because all brake shoe assemblies are constructed of more than one piece.
Answer B is wrong because shoes or linings are cast, not molded.
Answer C is wrong because brake shoes are not staked.
Answer D is correct because rivets or bolts are used to fasten the linings to the shoes.

Question #21
Answer A is a good choice because a hydraulic hydrobooster uses a balanced steering gear and can be integrated with the power steering pump. Yet, it is wrong because both technicians are right.
Answer B is also a good choice because a dedicated hydraulic power source not requiring a balanced steering gear can be used to power a hydraulic hydrobooster. Yet, it is wrong because both technicians are right.
Answer C is correct because of the above reasons and because both technicians are right.
Answer D is wrong because both technicians are right.

Question #22
Answer A is wrong because a visual inspection is not a required part of a governor cut-out test.
Answer B is correct because as part of the system check, you drain the system completely to zero.
Answer C is wrong because only one technician is right.
Answer D is wrong because one technician is right.

Question #23
Answer A is correct because the axle housing assembly supports the vehicle's unsprung weight.
Answer B is wrong because the axle supports the vehicle's weight on a semifloating axle.
Answer C is wrong because the wheel does not support the vehicle's weight.
Answer D is wrong because an inner axle bearing will support the weight of the differential.

Question #24
Answer A is wrong because the S-cam is positioned either in front of or behind the axle.
Answer B is wrong because the S-cam is positioned either in front of or behind the axle.
Answer C is wrong because wheel rotation dictates that the primary shoe be positioned at the bottom.
Answer D is correct because the primary shoe is the first shoe past the cam in the direction of wheel rotation. It goes on the bottom.

Question #25
Answer A is a good choice because you do use double flare tubing when replacing brake lines. Yet, it is wrong because both technicians are right.
Answer B is also a good choice because you do use ISO tubing when replacing a brake line. Yet, it is wrong because both technicians are right.
Answer C is correct because of the above reasons and because both technicians are right.
Answer D is wrong because both technicians are right.

Question #26
Answer A is wrong because a loose drive pulley belt will cause slipping and "squealing."
Answer B is wrong because a restriction in the unit will cause the compressor to make more "noise."
Answer C is correct because a defective head gasket is the least likely cause of slipping.
Answer D is wrong because improper lubrication will cause increased volume levels from high friction parts.

Question #27
Answer A is wrong because the unloader valve is not part of an air dryer assembly. It is part of the compressor.
Answer B is correct because a purge valve is used in an air dryer.
Answer C is wrong because the solenoid valve is not part of an air dryer assembly.
Answer D is wrong because the desiccant holder is not pictured in the figure.

Question #28
Answer A is correct because you do move the handle to the fully applied position and record the air pressure.
Answer B is wrong because the air system does not have to be drained to test the trailer control valve.
Answer C is wrong because as long as the trailer brakes are fully charged, the initial gauge reading is unnecessary.
Answer D is wrong because air leakage around the control handle should not occur.

Question #29
Answer A is wrong because contaminated linings will cause reduced braking power/force.
Answer B is wrong because a leaking quick-release valve will cause a reduction in air pressure.
Answer C is wrong because a slack adjuster orientation to the chamber pushrod is important to maximize full braking force.
Answer D is correct because a leaking power booster only affects assist not braking force.

Question #30
Answer A is wrong because ideal braking forces require proportionality, correct timing, and identical pressure levels.
Answer B is correct because it is one in which the braking pressure reaches each actuator simultaneously and at the same pressure level.
Answer C is wrong because wheels will lock up receiving pressure first if given in sequential order.
Answer D is wrong because varying load conditions will prohibit a "graduated" braking force, because of unequal braking force(s).

Question #31
Answer A is wrong because a damaged or defective chamber diaphragm will cause excessive leakage.
Answer B is wrong because a damaged hose or fitting will cause leakage when pressurized.
Answer C is correct because defective compressor gaskets will not cause brake system leakage.
Answer D is wrong because a defective relay valve will cause an excessive air leak.

Question #32
Answer A is a good choice because the applied stroke uses an 80 psi brake application. Yet, it is wrong because both technicians are right.
Answer B is also a good choice because the brakes are adjusted (if needed) to achieve the proper free stroke. Yet, it is wrong because both technicians are right.
Answer C is correct because of the above reasons and because both technicians are right.
Answer D is wrong because both technicians are right.

Question #33
Answer A is wrong because the manufacturer's procedures should always be followed.
Answer B is correct because you always follow the manufacturer's procedures.
Answer C is wrong because only one technician is right.
Answer D is wrong because one technician is right.

Question #34
Answer A is wrong because a caliper piston seized in its bore would usually wear both pads evenly.
Answer B is wrong because a damaged and/or defective rotor will usually wear both pads evenly.
Answer C is wrong because a damaged brake hose/line will tend to wear both pads evenly.
Answer D is correct because a caliper seized on the slide pins will not release or react to the braking force and wear the outboard pad.

Question #35
Answer A is a good choice because wheel and axle sensors are electromagnetic devices used to obtain vehicle speed information. Yet, it is wrong because both technicians are right.
Answer B is also a good choice because when an ABS system is disabled, the vehicle goes into its standard braking mode. Yet, it is wrong because both technicians are right.
Answer C is correct because of the above reasons and because both technicians are right.
Answer D is wrong because both technicians are right.

Question #36
Answer A is correct because excessive wear limits between the clevis and the collar will cause the stroke to be too long.
Answer B is wrong because an automatic slack adjuster needs no adjustment.
Answer C is wrong because only one technician is right.
Answer D is wrong because one technician is right.

Explanations to the Answers for the Sample Test Section 5

Question #37
Answer A is wrong because incompatible fluid will cause rubber cup seals in the wheel cylinders to swell and deteriorate.
Answer B is wrong because incompatibility with specific moisture rates will ensue when improper brake fluid is used in noncompatible systems.
Answer C is wrong because incompatibility with specific boil rates will be encountered when lower boil-rate fluids are used in heavier vehicles.
Answer D is correct because contamination of the air in air-over-hydraulic systems is the least likely effect of using the wrong brake fluid.

Question #38
Answer A is wrong because to facilitate removing the spring chamber from the adapter, you must hold the adapter while sliding the spring chamber sideways.
Answer B is wrong because you remove the diaphragm by turning it counterclockwise from the chamber.
Answer C is correct because you loosen the clamps by lightly striking them with a hammer or soft mallet.
Answer D is wrong because the air pressure should be exhausted if it was used to aid in caging the spring brake.

Question #39
Answer A is wrong because an aftercooler has a permanent filter that traps oil and particulate matter.
Answer B is correct because aftercooler air dryers employ no substances or materials to remove moisture.
Answer C is wrong because an aftercooler receives hot air from the compressor.
Answer D is wrong because an aftercooler dryer needs no daily maintenance.

Question #40
Answer A is correct because quick-release valves are located close to the brake chambers.
Answer B is wrong because quick-release valves are not mounted in the cab.
Answer C is wrong because quick-release valves must be mounted close to the chambers they serve.
Answer D is wrong because quick-release valves must be mounted close to the chambers they serve.

Question #41
Answer A is a good choice because a type 30 rotochamber has an effective diaphragm area of 30 square inches. Yet, it is wrong because both technicians are right.
Answer B is also a good choice because a rotochamber's operation is similar to the conventional type brake chamber. Yet, it is wrong because both technicians are right.
Answer C is correct because of the above reasons and because both technicians are right.
Answer D is wrong because both technicians are right.

Question #42
Answer A is correct because it is okay to machine oversize drums as long as the drum's manufacturer's recommendation for machining dimensions are followed.
Answer B is wrong because turning an oversize drum decreases its overall strength.
Answer C is wrong because only one technician is right.
Answer D is wrong because one technician is right.

Question #43
Answer A is wrong because a buzzer may be interfaced to a truck's power hydraulic brake booster.
Answer B is wrong because foundation brakes are interfaced with a truck's hydraulic brake booster.
Answer C is wrong because batteries interface with a truck's power hydraulic power booster.
Answer D is correct because the truck alternator will not typically interface with the brake booster.

Question #44
Answer A is wrong because a hydraulic wheel cylinder may occasionally need to be re-bored.
Answer B is correct because you always inspect wheel cylinders on a PM.
Answer C is wrong because a wheel cylinder does not need replacing periodically, but needs servicing on occasion.
Answer D is wrong because wheel cylinders are generally constructed of cast iron.

Question #45
Answer A is correct because the ECU will no longer cover that portion of the system.
Answer B is wrong because full functioning of the entire ABS system is not available when the ECU shuts down a portion of that system.
Answer C is wrong because the ECU will not completely shut down when a failure of only a portion of the ABS system has occurred.
Answer D is wrong because all defaults are instantly monitored by the ECU.

Question #46
Answer A is correct because if any slack adjuster defect is found, the slack adjuster is to be replaced.
Answer B is wrong because automatic slack adjusters do require lubrication.
Answer C is wrong because only one technician is right.
Answer D is wrong because one technician is right.

Question #47
Answer A is wrong because an air gauge is all that is required for the timing test.
Answer B is correct because balanced air timing means each axle receives air at the same time and at the same pressure level as needed by the driver.
Answer C is wrong because only one technician is right.
Answer D is wrong because one technician is right.

Question #48
Answer A is wrong because fixed, sliding, and floating calipers are about the same in price.
Answer B is wrong because fixed, sliding, and floating calipers are about the same in price.
Answer C is wrong because fixed, sliding, and floating calipers are generally the same in size.
Answer D is correct because a fixed caliper has pistons on each side of a brake rotor, while a floating or sliding caliper only has piston(s) on one side of a brake rotor.

Question #49
Answer A is a good choice because it states no attempt should be made to repair/replace any part of the piggyback assembly, but it should be completely replaced as a unit. Yet, it is wrong because both technicians are right.
Answer B is also a good choice because the spring chamber should be disarmed (manually released), before discarding. Yet, it is wrong because both technicians are right.
Answer C is correct because of the above reasons and because both technicians are right.
Answer D is wrong because both technicians are right.

Question #50
Answer A is correct because it does contain a separate inlet/exhaust valve that controls two separate reservoirs.
Answer B is wrong because a dual-circuit brake valve is a treadle-type, usually mounted on the floor.
Answer C is wrong because a trailer supply valve is a push-pull valve mounted on the instrument front panel.
Answer D is wrong because the dual circuit brake valve operates in a linear or straight line motion.

Appendices Explanations to the Answers for the Sample Test Section 5

Question #51
Answer A is a good choice because the duo-servo brake system uses different primary and secondary shoes. Yet, it is wrong because both technicians are right.
Answer B is also a good choice because the primary shoe in a servo-type brake system has a weaker return spring. Yet, it is wrong because both technicians are right.
Answer C is correct because of the above reasons and because both technicians are right.
Answer D is wrong because both technicians are right.

Question #52
Answer A is wrong because a seized brake caliper piston will not cause brake fade—only pad wear and reduced braking power.
Answer B is correct because a brake drum machined beyond its limit causes the shoes to move farther for contact. Brake drum diameter increases with heat when the drum is heated during a severe or prolonged brake application. When heat causes brake drum expansion, one must depress the brake pedal further to force the shoes against the drum. Industry calls this action BRAKE FADE.
Answer C is wrong because leakage past the master cylinder cup will cause a low or no pedal, not brake fade.
Answer D is correct because air in the hydraulic system will cause a spongy pedal, not brake fade.

Question #53
Answer A is wrong because the spring does not release, but applies.
Answer B is wrong because air begins to exhaust from the piggyback chamber.
Answer C is wrong because the treadle or application valve is not part of the park control valve circuit.
Answer D is correct because air in the rear portion of the spring brake chamber is exhausted and the springs apply the parking brakes.

Question #54
Answer A is a good choice because when wedge brakes are not properly adjusted, this reduces brake advantage and contributes to poor brake performance. Yet, it is wrong because both technicians are right.
Answer B is also a good choice because a properly adjusted brake is one where the angle between the slack adjuster and the chamber pushrod is 90° when the brakes are applied. Yet, it is wrong because both technicians are right.
Answer C is correct because of the above reasons and because both technicians are right.
Answer D is wrong because both technicians are right.

Question #55
Answer A is a good choice because the brakes may be wedge-actuated. Yet, it is wrong because both technicians are right.
Answer B is also a good choice because the brakes may also be cam actuated. Yet, it is wrong because both technicians are right.
Answer C is correct because of the above reasons and because both technicians are right.
Answer D is wrong because both technicians are right.

Question #56
Answer A is wrong because a pressure differential valve senses pressure.
Answer B is correct because holdoff and metering valve are the same.
Answer C is wrong because the load proportioning valve senses a vehicle's weight.
Answer D is wrong because a combination valve is a multifunction valve housing the pressure differential valve and metering valves in one housing.

Question #57
Answer A is a good choice because it uses the same sequence of riveting as on the shoe previously. Yet, it is wrong because both technicians are right.
Answer B is also a good choice because you always follow the manufacturer's procedures when riveting linings on shoes. Yet, it is wrong because both technicians are right.
Answer C is correct because of the above reasons and because both technicians are right.
Answer D is wrong because both technicians are right.

Question #58
Answer A is wrong because you use gladhands in an air system.
Answer B is wrong because you always use a proportioning valve in a front disc-rear drum non-ABS system.
Answer C is wrong because neither technician is right.
Answer D is correct because both technicians are right.

Question #59
Answer A is a good choice because a leaking diaphragm or seal could cause dragging brakes. Yet, it is wrong because both technicians are right.
Answer B is also a good choice because a broken return spring in the spring brake section could be the cause of dragging brakes. Yet, it is wrong because both technicians are right.
Answer C is correct because of the above reasons and because both technicians are right.
Answer D is wrong because both technicians are right.

Question #60
Answer A is correct because blocked compensation ports will cause brake drag.
Answer B is wrong because excessive pedal free play will not cause brake drag. Insufficient pedal free play will cause brake drag.
Answer C is wrong because only one technician is right.
Answer D is wrong because one technician is right.

Question #61
Answer A is correct because fluid contamination will result from leaving the reservoir cover off.
Answer B is wrong because loss of pressure will not result from leaving the reservoir cover off.
Answer C is wrong because seal erosion will not result from leaving the reservoir cover off.
Answer D is wrong because fluid aeration will not result from leaving the reservoir cover off.

Question #62
Answer A is correct because seized caliper pistons may cause brake drag.
Answer B is wrong because rotors machined beyond the maximum diameter may cause brake pedal fade when you first apply the brakes—not drag.
Answer C is wrong because only one technician is right.
Answer D is wrong because one technician is right.

Question #63
Answer A is wrong because a bell-mouth condition results when the drum diameter is greater at the edge of the drum next to the backing plate than the edge near the wheel.
Answer B is wrong because a concave brake drum results when the drum diameter is greater in the center than that of both edges.
Answer C is wrong because an out-of-round condition results when there is a variation in readings 180° apart.
Answer D is correct because a convex drum results when the diameter on a driveline parking brake drum is greater at the edges of the friction surface than in the center.

Question #64
Answer A is wrong because low fluid level will not cause brake fade.
Answer B is correct because brake drums worn or machined beyond the maximum diameter could cause brake fade.
Answer C is wrong because only one technician is right.
Answer D is wrong because one technician is right.

Question #65
Answer A is correct because you must remove the wheel in most cases to measure pad thickness.
Answer B is wrong because if the brakes were applied shortly before measuring the lining to rotor clearance, this clearance will be LESS than specified.
Answer C is wrong because only one technician is right.
Answer D is wrong because one technician is right.

Question #66
Answer A is a good choice because a one-way check valve that is restricted in the vacuum hose to the hydrovac unit will cause excessive brake effort. Yet, it is wrong because both technicians are right.
Answer B is also a good choice because it states that a sticking power piston in the hydrovac unit could also cause excessive brake effort. Yet, it is wrong because both technicians are right.
Answer C is correct because of the above reasons because both technicians are right.
Answer D is wrong because both technicians are right.

Question #67
Answer A is wrong because diagonally split systems refer to the service system, not the parking system.
Answer B is wrong because air-operated systems have spring brakes for their parking systems.
Answer C is wrong because some hydraulic brake systems have spring brakes for parking.
Answer D is correct because some air-over-hydraulic parking systems use cable-operated parking brakes.

Answers to the Test Questions for the Additional Test Questions Section 6

1.	B	31.	A	61.	D	91.	B
2.	D	32.	D	62.	B	92.	B
3.	C	33.	B	63.	A	93.	D
4.	A	34.	C	64.	D	94.	D
5.	D	35.	B	65.	C	95.	C
6.	C	36.	A	66.	A	96.	D
7.	C	37.	D	67.	A	97.	B
8.	A	38.	C	68.	A	98.	B
9.	B	39.	D	69.	C	99.	D
10.	A	40.	B	70.	D	100.	B
11.	D	41.	A	71.	C	101.	B
12.	B	42.	C	72.	D	102.	C
13.	A	43.	A	73.	A	103.	C
14.	D	44.	B	74.	A	104.	D
15.	A	45.	D	75.	A	105.	B
16.	B	46.	A	76.	B	106.	A
17.	A	47.	D	77.	A	107.	A
18.	C	48.	D	78.	B	108.	C
19.	D	49.	A	79.	A	109.	A
20.	C	50.	C	80.	B	110.	A
21.	A	51.	B	81.	C	111.	A
22.	B	52.	C	82.	D	112.	B
23.	A	53.	C	83.	A	113.	C
24.	D	54.	A	84.	C	114.	D
25.	C	55.	B	85.	D	115.	C
26.	D	56.	D	86.	B	116.	A
27.	D	57.	A	87.	A	117.	D
28.	C	58.	C	88.	A	118.	C
29.	B	59.	B	89.	A	119.	D
30.	B	60.	B	90.	B	120.	B

Explanations to the Answers for the Additional Test Questions Section 6

Question #1
Answer A is wrong because a one-way check valve ensures that air flows in one direction at one time.
Answer B is correct because if one reservoir loses pressure, the double check valve alerts the functioning reservoir to supply air to both circuits.
Answer C is wrong because a double check valve does not serve as a check valve.
Answer D is wrong because a re-settable spring is responsible for governor cut-off levels.

Question #2
Answer A is wrong because 100 psi is not necessary for a brake chamber application.
Answer B is wrong because 80 psi is not necessary for a brake chamber application.
Answer C is wrong because 40 psi is not necessary for a brake chamber application.
Answer D is correct because 20 psi or less is necessary.

Question #3
Answer A is a good choice because the park and emergency braking system is a separate and distinct air circuit, completely isolated from the regular service air system. Yet, it is wrong because both technicians are right.
Answer B is also a good choice because spring brake chambers use air pressure in the opposite way from the service brake chambers. Yet, it is wrong because both technicians are right.
Answer C is correct because of the above reasons and because both technicians are right.
Answer D is wrong because both technicians are right.

Question #4
Answer A is correct because you use high-quality, high-temperature grease when repacking is necessary.
Answer B is wrong because a wheel bearing is not transferable from one bearing race to another.
Answer C is wrong because only one technician is right.
Answer D is wrong because one technician is right.

Question #5
Answer A is wrong because reversing the gladhand positions only reverses the switching for those valves.
Answer B is wrong because a tractor "running bobtail" is a tractor running without a trailer.
Answer C is wrong because the quick-release valve releases air from a chamber rapidly.
Answer D is correct because simultaneous application of the service and emergency side of the spring brake chambers is anticompounding.

Question #6
Answer A is a good choice because you do install the lip of the seal towards the outside of the adjuster. Yet, it is wrong because both technicians are right.
Answer B is also a good choice because you do not hit the seal after it reaches the bottom of the bore. Yet, it is wrong because both technicians are right.
Answer C is correct because of the above reasons and because both technicians are right.
Answer D is wrong because both technicians are right.

Question #7
Answer A is a good choice because power brake boosters can be found on hydraulically controlled brake systems. Yet, it is wrong because both technicians are right.
Answer B is also a good choice because power brake boosters are found on air-over-hydraulic systems. Yet, it is wrong because both technicians are right.
Answer C is correct because of the above reasons and because both technicians are right.
Answer D is wrong because both technicians are right.

Question #8
Answer A is correct because cam-actuated brake systems are the most popular brake system on heavy duty trucks today.
Answer B is wrong because fabricated brake shoes use an integral pin and roller design for their contact points.
Answer C is wrong because only one technician is right.
Answer D is wrong because one technician is right.

Question #9
Answer A is wrong because the discharge valve does not manage proper air direction into the wet tank.
Answer B is correct because the discharge valve prevents the compressed air in the discharge line from returning to the cylinder bore as the intake and compression cycle is repeated.
Answer C is wrong because the discharge valve does not provide pressure relief.
Answer D is wrong because the discharge valve allows air to pass into the discharge line.

Question #10
Answer A is correct because gladhand seals are made from rubber.
Answer B is wrong because polyurethane is not as widely used as rubber for gladhand seals.
Answer C is wrong because plastic is not generally used for gladhand seals.
Answer D is wrong because gladhand seals are not made from rubber-covered aluminum.

Question #11
Answer A is wrong because "warning valve" is common, but not "pressure warning valve."
Answer B is wrong because "system inproportional indicator" is not a name for the pressure differential valve.
Answer C is wrong because "pressure valve" is not a common name for the pressure differential valve.
Answer D is correct because "pressure warning valve" is the common name for the pressure differential valve.

Question #12
Answer A is wrong because a heavy duty truck with an ABS has one modulator per wheel.
Answer B is correct because a truck's ABS brake system is designed to handle a split coefficient situation when that occurs.
Answer C is wrong because only one technician is right.
Answer D is wrong because one technician is right.

Question #13
Answer A is correct because fabricated shoes are most common.
Answer B is wrong because cast brake shoes are not very common.
Answer C is wrong because cam-actuation is a description of energization, not of construction.
Answer D is wrong because wedge-actuation is a description of energization, not of construction.

Question #14
Answer A is wrong because two ports are insufficient for a tandem axle.
Answer B is wrong because four ports are insufficient for a tandem axle.
Answer C is wrong because five ports are insufficient for a tandem axle.
Answer D is correct because seven are sufficient for a tandem axle.

Question #15
Answer A is correct because when the primary or secondary system fails, federal law requires that the parking brakes must release at least once.
Answer B is wrong because the parking brake will apply when air is released from the spring-brake chamber.
Answer C is wrong because only one technician is right.
Answer D is wrong because one technician is right.

Appendices   Explanations to the Answers for the Additional Test Questions Section 6   105

Question #16
Answer A is wrong because the chamber pushrod is not the final link, but the second-to-final link.
Answer B is correct because the slack adjuster is the final link.
Answer C is wrong because the camshaft is considered part of the foundation brake.
Answer D is wrong because the chamber mounting plate serves only to mount the foundation brake to the axle.

Question #17
Answer A is correct because it is friction built up in/on drum/rotor that actually stops vehicle motion.
Answer B is wrong because friction stops a vehicle, not the shoes which are the instruments needed to create friction.
Answer C is wrong because only one technician is right.
Answer D is wrong because one technician is right.

Question #18
Answer A is a good choice because one of the factors affecting air timing is the size of the lines. Yet, it is wrong because both technicians are right.
Answer B is also a good choice because if a trailer service system has an air fitting that is the wrong application, then the air timing/balance will be affected. Yet, it is wrong because both technicians are right.
Answer C is correct because of the above reasons and because both technicians are right.
Answer D is wrong because both technicians are right.

Question #19
Answer A is wrong because a hydraulic pressure gauge is used to test master cylinder output.
Answer B is wrong because wheel cylinders are generally rebuilt, not replaced.
Answer C is wrong because both technicians are wrong.
Answer D is correct because neither technician is right.

Question #20
Answer A is a good choice because a slack adjuster must be adjusted so the angle between it and the chamber pushrod is 90° when the brakes are fully applied. Yet, it is wrong because both technicians are right.
Answer B is also a good choice because the chamber pushrod should be adjusted with the shortest possible stroke, without dragging the brakes. Yet, it is wrong because both technicians are right.
Answer C is correct because of the above reasons and because both technicians are right.
Answer D is wrong because both technicians are right.

Question #21
Answer A is correct because overloading is the least likely cause of high heat buildup.
Answer B is wrong because contamination will cause high heat build up, destroying the bearing.
Answer C is wrong because axle housing damage may cause wheel bearings to fail.
Answer D is wrong because improper or incompatible lubricant will cause failure of the bearing.

Question #22
Answer A is wrong because gladhands seals ensure leakage at the gladhands connections to the trailer.
Answer B is correct because gladhands provides a means of preserving air pressure in sufficient amount to stop the tractor in the event of a trailer breakaway.
Answer C is wrong because the tractor protection valve does not hold off the spring brake in case of sudden air loss in the spring brake chamber.
Answer D is wrong because the bobtail proportioning valve senses when the tractor is running trailerless.

Question #23
Answer A is correct because power boosters can have dual diaphragms.
Answer B is wrong because the power brake booster "assists" in converting mechanical pressure into hydraulic pressure.
Answer C is wrong because most boosters are repairable as needed.
Answer D is wrong because Boosters are generally not exchangeable with each other.

Question #24
Answer A is wrong because Water should not be used because of its corrosive capabilities.
Answer B is wrong because mineral spirits have corrosive capabilities and are incompatible with some materials.
Answer C is wrong because gasoline should not be used because of its corrosiveness and incompatibilities with some materials.
Answer D is correct because alcohol is the most appropriate cleaning agent for flushing a hydraulic brake system.

Question #25
Answer A is a good choice because the proportioning valve should be inspected every time the brakes are serviced. Yet, it is wrong because both technicians are right.
Answer B is also a good choice because you use gauges ahead of and behind the proportioning valve to test its function. Yet, it is wrong because both technicians are right.
Answer C is correct because of the above reasons and because both technicians are right.
Answer D is wrong because both technicians are right.

Question #26
Answer A is wrong because 0.010–0.020 inch is not an acceptable clearance.
Answer B is correct because 0.008–0.016 inch is acceptable.
Answer C is wrong because 0.005–0.010 inch is not an acceptable clearance.
Answer D is wrong because 0.004–0.012 inch is not an acceptable clearance.

Question #27
Answer A is wrong because diagonally split systems refer to the service system, not the parking system.
Answer B is wrong because air-operated systems have spring brakes for their parking systems.
Answer C is wrong because some hydraulic brake systems have spring brakes for parking.
Answer D is correct because some air-over-hydraulic parking systems use spring brakes.

Question #28
Answer A is a good choice because if normal flow from the hydraulic brake pump is interrupted, the electric motor pump provides power for reserve stops. Yet, it is wrong because both technicians are right.
Answer B is also a good choice because manual braking is possible, if the power and reserve systems fail, but the braking will be extremely difficult and only emergency use of the vehicle is recommended. Yet, it is wrong because both technicians are right.
Answer C is correct because of the above reasons and because both technicians are right.
Answer D is wrong because both technicians are right.

Question #29
Answer A is wrong because an electronic braking system has an electronic computer.
Answer B is correct because an integrated braking system ensures that the correct brake pressure is distributed to each wheel at any given moment.
Answer C is wrong because an electronic brake control does not use digital relays.
Answer D is wrong because an electronic brake signal travels to the computer first, then to the components.

Question #30
Answer A is wrong because a micrometer is used to check thickness variation in brake rotors.
Answer B is correct because the caliper pistons are responsible for stopping the vehicle by forcing the pads against the rotors.
Answer C is wrong because only one technician is right.
Answer D is wrong because one technician is right.

Question #31
Answer A is correct because there are more valves between the foot valve and the trailer axles than the tractor axles, thus increasing restrictions and lowering those respective pressures.
Answer B is wrong because the increased length of air line increases the time for air delivery.
Answer C is wrong because the combination vehicle service lines are integrated via relay valves.
Answer D is wrong because the treadle valve operates both tractor and trailer air systems.

Question #32
Answer A is wrong because pneumatic windshield wipers can be driven by an air compressor.
Answer B is wrong because steering units can be driven by an air compressor.
Answer C is wrong because a fuel or steering pump can be driven by an air compressor.
Answer D is correct because providing air for the heater is the least likely use for air.

Question #33
Answer A is wrong because wedge-type brake systems are self-adjusting.
Answer B is correct because the proper free stroke is the reason for adjusting drum-to-lining clearance on cam-type slack adjusting brake systems.
Answer C is wrong because pushrod length is not as important a measurement as drum-to-lining clearance.
Answer D is wrong because freestroke is more critical than pushrod to slack adjuster orientation (90° angle).

Question #34
Answer A is a good choice because the supply reservoir is called a wet tank because moisture forms in this tank as the compressed air cools down. Yet, it is wrong because both technicians are right.
Answer B is also a good choice because in a two-reservoir system (one supply and one service), the two reservoirs work independently of each other; thus, if pressure in one reservoir fails, the second reservoir remains operable. Yet, it is wrong because both technicians are right.
Answer C is correct because of the above reasons and because both technicians are right.
Answer D is wrong because both technicians are right.

Question #35
Answer A is wrong because receiving the proper lubrication will not overcome overtorquing a bearing.
Answer B is correct because nothing happens as long as the initial procedure of torquing and correct torque values were followed according to the manufacturer's specifications.
Answer C is wrong because a wheel bearing seats itself immediately after the wheel nut is backed off.
Answer D is wrong because the adjusting nut must be backed off in order for the bearing to properly seat itself.

Question #36
Answer A is correct because you replace all the pads at the same time.
Answer B is wrong because brake pads should not be replaced on one side only at any one time.
Answer C is wrong because brake pads should always be changed in pairs.
Answer D is wrong because resurfacing of a brake rotor when pads are replaced is at the technician's discretion.

Question #37
Answer A is wrong because different size linings are not criteria for a definition of combination linings.
Answer B is wrong because differences in construction are not criteria for a definition of combination linings.
Answer C is wrong because both technicians are wrong.
Answer D is correct because neither technician is right.

Question #38
Answer A is wrong because glass is not generally used for containing brake fluid.
Answer B is wrong because all containers must be tightly sealed when storing brake fluid.
Answer C is correct because you never use any container that has been used to store any other liquid.
Answer D is wrong because brake fluid from a tightly sealed container may be used after one year of storage.

Question #39
Answer A is wrong because the S-cam is part of the slack adjuster.
Answer B is wrong because a push plate is used in the construction of a spring brake chamber.
Answer C is wrong because a check valve is part of the service air line system.
Answer D is correct because a pressure plate is shown in the figure.

Question #40
Answer A is wrong because a typical wheel sensor generates an AC signal.
Answer B is correct because the waveform signal generated is sinusoidal and is proportional to wheel speed.
Answer C is wrong because only one technician is right.
Answer D is wrong because one technician is right.

Question #41
Answer A is correct because the induction of clean air is the most important aspect of compressor maintenance.
Answer B is wrong because the air compressor shares the same oil with the vehicle engine.
Answer C is wrong because an inspection for a properly mounted compressor is considered a technician's discretion.
Answer D is wrong because contaminated air more profoundly affects the air system than governor performance.

Question #42
Answer A is a good choice because a hydraulic disc brake caliper can have two pistons. Yet, it is wrong because both technicians are right.
Answer B is also a good choice because a disc brake caliper can have four pistons. Yet, it is wrong because both technicians are right.
Answer C is correct because of the above reasons and because both technicians are right.
Answer D is wrong because both technicians are right.

Question #43
Answer A is correct because the line must be a hydraulic hose conforming to SAE J189.
Answer B is wrong because a hydro-boost brake system is bled by turning over the vehicle engine without starting it, and refilling accordingly with fluid.
Answer C is wrong because only one technician is right.
Answer D is wrong because one technician is right.

Question #44
Answer A is wrong because 2 psi is not the accepted industry limit for differences in combination air pressures.
Answer B is correct because 4 psi is the industry standard.
Answer C is wrong because 6 psi is not the accepted industry limit for differences in combination air pressures.
Answer D is wrong because 8 psi is not the accepted industry limit for differences in combination air pressures.

Question #45
Answer A is wrong because full-clevis free stroke is not a type of adjustment or setting.
Answer B is wrong because half-clevis free stroke is not a type of adjustment or setting.
Answer C is wrong because full free stroke is not a type of adjustment or setting.
Answer D is correct because initial free stroke is the correct adjustment term.

Question #46
Answer A is correct because both front- and rear-wheel bearings are adjustable.
Answer B is wrong because a brass drift punch is not needed to remove an old wheel bearing.
Answer C is wrong because only one technician is right.
Answer D is wrong because one technician is right.

Question #47
Answer A is wrong because servo action is used to define the dynamics of brake shoes on a brake drum.
Answer B is wrong because nonservo action is used to define the dynamics of brake shoes on a brake drum.
Answer C is wrong because energizing action is a term used to describe how a brake shoe is pulled tighter against the inside of a rotating drum. It does not apply to disc brakes.
Answer D is correct because action-reaction describes the operation of single piston disc brakes.

Question #48
Answer A is wrong because check valves regulate the return flow of brake fluid to the master cylinder.
Answer B is wrong because the metering valve prevents front application of the disc brakes to allow the rear wheel cylinders to build up equal pressure.
Answer C is wrong because sliding rubber seals in the wheel cylinders contain the brake fluid and prevent leakage.
Answer D is correct because the return spring as shown will return the brake pedal to its original position.

Question #49
Answer A is correct because the pushrod spring, in combination with the brake shoe return spring, releases the brakes.
Answer B is wrong because the clevis or yoke threads onto the brake chamber pushrod, which is also threaded to receive it.
Answer C is wrong because only one technician is right.
Answer D is wrong because one technician is right.

Question #50
Answer A is a good choice because on brakes with vented discs, the inboard pad could be thicker than the outboard pad. Yet, it is wrong because both technicians are right.
Answer B is also a good choice because on brakes with solid discs, the inboard and outboard pads could be the same thickness. Yet, it is wrong because both technicians are right.
Answer C is correct because of the above reasons and because both technicians are right.
Answer D is wrong because both technicians are right.

Question #51
Answer A is wrong because governors do not have check valves.
Answer B is correct because too much clearance at the compressor unloading valves could cause higher than normal air pressure.
Answer C is wrong because only one technician is right.
Answer D is wrong because one technician is right.

Question #52
Answer A is a good choice because a dual-power hydraulic brake booster could have one diaphragm. Yet, it is wrong because both technicians are right.
Answer B is also a good choice because a dual-power hydraulic brake booster could have two diaphragms. Yet, it is wrong because both technicians are right.
Answer C is correct because of the above reasons and because both technicians are right.
Answer D is wrong because both technicians are right.

Question #53
Answer A is a good choice because any time a wheel with a unitized seal is removed, the seal **must** be replaced. Yet, it is wrong because both technicians are right.
Answer B is also a good choice because wheel seals should be removed using a brass drift and a hammer. Yet, it is wrong because both technicians are right.
Answer C is correct because of the above reasons and because both technicians are right.
Answer D is wrong because both technicians are right.

Question #54
Answer A is correct because 85–100 psi in 25 seconds or less is the acceptable specification.
Answer B is wrong because 60–80 psi in 20 seconds is not the generally accepted standard for compressor performance.
Answer C is wrong because 90–110 psi in 30 seconds is not the generally accepted standard for compressor performance.
Answer D is wrong because 50–110 psi in 25 seconds is not the generally accepted standard for compressor performance.

Question #55
Answer A is correct because an improperly mounted axle is the least likely cause of brake torque imbalance.
Answer B is wrong because a stuck relay valve will cause braking problems.
Answer C is wrong because a broken shoe return spring will cause uneven braking.
Answer D is wrong because a brake chamber diaphragm failure will cause uneven braking.

Question #56
Answer A is correct because when performing a brake overhaul on a hydraulic rear drum brake system, you replace the return springs.
Answer B is wrong because the maximum allowable out-of-round specification on a brake drum is 0.015 inch.
Answer C is wrong because only one technician is right.
Answer D is wrong because one technician is right.

Question #57
Answer A is correct because the application valve shown has two pistons: a reaction or rear-modulating piston, and a primary or front-modulating piston.
Answer B is wrong because the reaction spring between the two pistons of a brake valve is NOT responsible for the basic operation of the valve.
Answer C is wrong because only one technician is right.
Answer D is wrong because one technicians are right.

Question #58
Answer A is wrong because the maximum discharge air temperature is 400°F.
Answer B is correct because most air compressors are liquid cooled.
Answer C is wrong because only one technician is right.
Answer D is wrong because one technician is right.

Question #59
Answer A is wrong because on disc wheels, the hub is a separate component assembly.
Answer B is correct because a wheel bearing may be locked in place by either an adjusting and lock nut assembly, split forged nut, or castellated nut and cotter pin assembly.
Answer C is wrong because only one technician is right.
Answer D is wrong because one technician is right.

Question #60
Answer A is correct because the air pressure at the brake application valve delivery port must be proportional to treadle movement.
Answer B is wrong because with the brakes applied, a 1-inch (2.54 cm) bubble in 3 seconds at the brake application valve exhaust port indicates normal leakage.
Answer C is wrong because only one technician is right.
Answer D is wrong because one technician is right.

Question #61
Answer A is wrong because you never wash master cylinder parts in an oil-based solvent.
Answer B is wrong because you never hone an aluminum master cylinder even if it is scored.
Answer C is wrong because both technicians are wrong.
Answer D is correct because neither technician is right.

Question #62
Answer A is wrong because with the service brakes applied, acceptable leakage at the quick-release valve exhaust port is a 1-inch (2.54 cm) bubble in 3 seconds.
Answer B is correct because when the service brakes are applied and released, the air pressure in the front brake chambers should be released immediately.
Answer C is wrong because only one technician is right.
Answer D is wrong because one technician is right.

Question #63
Answer A is correct because a clogged compressor inlet filter is the most likely cause of slow buildup.
Answer B is wrong because a leak in a brake chamber is not as likely of a cause for slow buildup.
Answer C is wrong because an air leak in the cab suspension is an unlikely cause.
Answer D is wrong because a restriction in the governor line will cause high pressure.

Question #64
Answer A is wrong because it opens when the reservoir pressure reaches 2 psi higher than reservoir pressure.
Answer B is wrong because it is mounted in a threaded fitting near the bottom of the reservoir.
Answer C is wrong because it opens by increasing supply reservoir pressure.
Answer D is correct because it drains water if the sump pressure is 2 psi higher than reservoir pressure.

Question #65
Answer A is wrong because decreasing tension to original specification level and recheck contact surface for excessive wear will not cure the condition.
Answer B is wrong because measuring the level of deflection at a section of belt that is longest between pulleys will not solve the problem.
Answer C is correct because you replace the belt if you notice contact at the bottom of the pulley with the belt.
Answer D is wrong because if you ignore it the problem will persist.

Question #66
Answer A is wrong because it is NOT necessary to retime the engine when this replacement process is complete.
Answer B is correct because you do inspect the gear for worn or chipped teeth.
Answer C is wrong because only one technician is right.
Answer D is wrong because one technician is right.

Question #67
Answer A is correct because you put a light-coating lubricant on the spindle and place a seal protector over the threads.
Answer B is wrong because rear wheel hub assemblies do use oil filler plugs to fill the wheel cavities.
Answer C is wrong because only one technician is right.
Answer D is wrong because one technician is right.

Question #68
Answer A is correct because the governor cut-out pressure may be increased by rotating the adjusting screw counterclockwise.
Answer B is wrong because you cannot adjust cut-in pressure.
Answer C is wrong because you turn the adjustment screw one-fourth turn to change cut-in or cut-out pressure 4 psi, not 7.
Answer D is wrong because if you restrict the airline from the supply reservoir to the governor, the cut-out pressure will increase.

Question #69
Answer A is a good choice because daily draining of air reservoirs is highly recommended. Yet, it is wrong because both technicians are right.
Answer B is also a good choice because you should check all automatic drain valves and moisture-removing devices periodically for proper operation. Yet, it is wrong because both technicians are right.
Answer C is correct because of the above reasons and because both technicians are right.
Answer D is wrong because both technicians are right.

Question #70
Answer A is wrong because a damaged airline may introduce moisture into the system.
Answer B is wrong because improper maintenance schedules on the drainage of supply tanks are very important in prohibiting moisture buildup on their interiors.
Answer C is wrong because damage to the supply tanks may cause moisture directly inside the affected area.
Answer D is correct because humid climate operation is the least likely cause of moisture in the system.

Question #71
Answer A is correct because you do move the handle to the fully applied position and record the air pressure.
Answer B is wrong because the air system does not have to be drained to test the trailer control valve.
Answer C is wrong because as long as the trailer brakes are fully charged, the initial gauge reading is unnecessary.
Answer D is wrong because air leakage around the control handle should not occur.

Question #72
Answer A is a good choice because a defective trailer control valve could be the cause of improper trailer brake operation. Yet, it is wrong because both technicians are right.
Answer B is also a good choice because excessive compressor cycling is a symptom of this same condition. Yet, it is wrong because both technicians are right.
Answer C is correct because of the above reasons and because both technicians are right.
Answer D is wrong because both technicians are right.

Question #73
Answer A is wrong because the inlet valve leakage is tested with the service brakes released.
Answer B is wrong because you do apply a soap solution to the area around the inlet and exhaust valve retaining ring to check exhaust valve leakage.
Answer C is wrong because exhaust valve leakage is tested with the brakes applied.
Answer D is correct because the control port on the service brake relay valve is connected to the delivery port on the brake application valve.

Question #74
Answer A is correct because air lines are connected from the service brake relay valve delivery ports to the rear axle service brake chambers.
Answer B is wrong because when the brakes are released, the inlet valve is closed, not open, in the service brake relay valve.
Answer C is wrong because when the brakes are released, the exhaust valve is open, not closed, in the service brake relay valve.
Answer D is wrong because the service brake relay valve is in a balanced position when the air pressure in the rear brake chambers equals system pressure.

Question #75
Answer A is correct because with the control valve in the dry road position, the limiting quick-release valve does not reduce air pressure to the front brakes.
Answer B is wrong because with the control valve in the slippery road position and the service brakes applied, leakage at the control valve exhaust port should not exceed a 1-inch (2.54 cm) bubble in 3 seconds.
Answer C is wrong because with the control valve in the slippery road position and the service brakes applied, leakage at the limiting quick-release valve exhaust port should not exceed a 1-inch (2.54 cm) bubble in 3 seconds.
Answer D is wrong because with the control valve in the slippery road position and the service brakes applied, it limits front brake pressure to 50 percent of the application valve pressure.

Question #76
Answer A is wrong because a bad tractor protection valve will not cause wheel lockup during a brake application.
Answer B is correct because a defective quick-release valve can cause slow front pressure buildup and poor stopping ability.
Answer C is wrong because only one technician is right.
Answer D is wrong because one technician is right.

Question #77
Answer A is correct because when you supply 50 psi (344.75 Kpa) from the brake application valve to the limiting valve inlet port, the pressure on a gauge at the limiting valve outlet port should be 40 psi (275.8 Kpa).
Answer B is wrong because 15 psi (103.42 Kpa) is the wrong value.
Answer C is wrong because 25 psi (172.37 Kpa) is the wrong value.
Answer D is wrong because 30 psi (206.85 Kpa) is the wrong value.

Question #78
Answer A is wrong because if the rear axle reservoir is drained to zero air pressure and the brake application valve is applied, the tractor air brake pressure should be at what ever is left in the front reservoir.
Answer B is correct because when the tractor parking brakes are applied, the air pressure at the spring brake control valve delivery port should drop quickly to zero. This zero air pressure allows the springs in the parking brakes to apply the brakes.
Answer C is wrong because only one technician is right.
Answer D is wrong because one technician is right.

Question #79
Answer A is correct because single check valves are connected between the primary and secondary reservoirs.
Answer B is wrong because a double check valve allows air pressure to flow from to different locations.
Answer C is wrong because only one technician is right.
Answer D is wrong because one technician is wrong.

Question #80
Answer A is wrong because a red low-pressure warning light on a vehicle will go out after startup even with a loss of one section of the dual system.
Answer B is correct because the problem may be caused by a loss of the complete air brake system.
Answer C is wrong because only one technician is right.
Answer D is wrong because one technician is right.

Question #81
Answer A is a good choice because an inoperative warning light and buzzer below 60 psi can be caused by a defective bulb. Yet, it is wrong because both technicians are right.
Answer B is also a good choice because an inaccurate dash pressure gauge can be the cause of a warning light on above-60 psi supply pressure. Yet, it is wrong because both technicians are right.
Answer C is correct because both technicians are right.
Answer D is wrong because both technicians are right.

Question #82
Answer A is wrong because grease on the linings on one axle will cause a reduced braking force on that axle.
Answer B is wrong because torque imbalance will result from the adjusting of one wheel tighter than another.
Answer C is wrong because mismatched sizing on service chambers will cause torque imbalance.
Answer D is correct because an overloaded vehicle will not cause torque imbalance or one wheel being adjusted tighter than another.

Question #83
Answer A is correct because excessive clearance between the linings and the brake drum may cause the cam to roll over during a brake application.
Answer B is wrong because this condition will not cause loss of steering control.
Answer C is wrong because only one technician is right.
Answer D is wrong because one technician is right.

Question #84
Answer A is wrong because both camshaft seal lips must face toward the slack adjuster.
Answer B is wrong because you do use the proper drive when installing camshaft seals.
Answer C is correct because you never lubricate the cam head with petroleum jelly or anything.
Answer D is wrong because you do measure camshaft radial movement.

Question #85
Answer A is wrong because installing a new wedge with a different angle will affect brake operation by changing the distance the brake shoes travel.
Answer B is wrong because you put lubricant on the wedge or wedge ramps.
Answer C is wrong because both technicians are wrong.
Answer D is correct because both technicians are wrong.

Question #86
Answer A is wrong because minor scoring may be removed from the actuator bores with fine emery cloth.
Answer B is correct because the threaded actuator opening is chamfered.
Answer C is wrong because the clearance between each wedge brake shoe lining and the drum should be 0.020 to 0.040 inch (0.51 to 1.02 mm).
Answer D is wrong because if the threaded actuator opening is chamfered, the taper on this collet nut faces the taper on the actuator opening.

Question #87
Answer A is correct because hydraulic hoses are not part of any air system.
Answer B is wrong because a torque plate is part of an air disc brake system.
Answer C is wrong because a caliper is part of an air disc brake system.
Answer D is wrong because a rotor is part of an air disc brake system.

Question #88
Answer A is correct because you do use the same procedure with bolted linings as with riveted ones.
Answer B is wrong because backing plate platforms on heavy duty trucks are not usually greased.
Answer C is wrong because only one technician is right.
Answer D is wrong because one technician is right.

Question #89
Answer A is correct because severe heat cracking will cause rapid lining wear.
Answer B is wrong because brake drums over 14 inches in diameter should not be increased more than 0.080 inch.
Answer C is wrong because only one technician is right.
Answer D is wrong because one technician is right.

Question #90
Answer A is wrong because high thickness variation does not cause reduced braking power.
Answer B is correct because high rotor thickness variation will cause pedal pulsation.
Answer C is wrong because high thickness variation will not cause loss of brake fluid.
Answer D is wrong because loss of directional control is not generally evident with high thickness variation.

Question #91
Answer A is wrong because an incorrect pushrod-to-slack adjuster arm angle will result in reduction of parking brake force.
Answer B is correct because a defective self-adjusting star wheel is the least likely cause of a parking brake that will not hold.
Answer C is wrong because a foundation brake problem will affect the spring parking brake system.
Answer D is wrong because weak spring brake chamber springs will reduce force on the slack adjusters.

Question #92
Answer A is wrong because "apply line" is not the common term for the control line.
Answer B is correct because "emergency line" is the common term for the control line.
Answer C is wrong because "park line" is not the common term for the control line.
Answer D is wrong because "trailer line" is not the common term for the control line.

Question #93
Answer A is wrong because the tractor park valve has a blue, round-shaped knob that is pulled out to apply the tractor parking brakes.
Answer B is wrong because the tractor park valve cannot be used to apply the trailer parking brakes.
Answer C is wrong because neither technician is right.
Answer D is correct because both technicians are wrong.

Question #94
Answer A is wrong because the trailer air supply valve allows air to enter the trailer brake system when the knob is pushed inward.
Answer B is wrong because the knob of the park control (system park) valve must be pushed in before the trailer air supply valve can receive air at the control port.
Answer C is wrong because neither technician is right.
Answer D is correct because both technicians are wrong.

Question #95
Answer A is a good choice because pressure below 45 psi in the air brake system will cause the knob of the trailer air supply valve to pop out. Yet, it is wrong because both technicians are right.
Answer B is also a good choice because pressure below 45 psi in the trailer air brake system will cause the control valve inside the tractor protection valve to close. Yet, it is wrong because both technicians are right.
Answer C is correct because both technicians are right.
Answer D is wrong because both technicians are right.

Question #96
Answer A is wrong because when the trailer air supply valve knob is pulled outward and the park control (system park) valve knob is pushed inward, the trailer air supply line is exhausted.
Answer B is wrong because when the trailer air supply valve knob is pulled outward and the park control (system park) valve knob is pushed inward, the tractor parking brakes are released.
Answer C is wrong because when the trailer air supply valve knob is pulled outward and the park control (system park) valve knob is pushed inward, air is exhausted through the trailer control valve.
Answer D is correct because when the trailer air supply valve knob is pulled outward and the park control (system park) valve knob is pushed inward, the trailer parking brakes are applied.

Question #97
Answer A is wrong because shorted solenoid coils will cause a default in the ABS circuit.
Answer B is correct because leaking hydraulic lines can cause a loss of pressure at a wheel with a resultant change in wheel speed.
Answer C is wrong because if an exhaust solenoid port is blocked, pressure will build in the system, initiating a fault.
Answer D is wrong because an open ECU will cause an ABS system shutdown.

Question #98
Answer A is wrong because all trucks over 10,000 lbs. GVW are not required to have dual action braking.
Answer B is correct because all new trucks, trailers, and heavy tractors built after 1996 must be equipped with an antilock braking system.
Answer C is wrong because type 35 rotochambers are not required on trucks over 20,000 lbs. GVW.
Answer D is wrong because forged steel brake chamber pushrods are not required on heavy trucks.

Question #99
Answer A is wrong because currently, there is no designed eight-channel system.
Answer B is wrong because a six-channel system is the second most widely used ABS system used today.
Answer C is wrong because the four-channel system is the most popular system in use today.
Answer D is correct because the two-channel system is the least used because of its limited control.

Question #100
Answer A is wrong because a seized brake caliper piston will not cause brake fade—only pad wear and reduced braking power.
Answer B is correct because a brake drum machined beyond its limit causes the shoes to move farther for contact. Brake drum diameter increases with heat when the drum is heated during a severe or prolonged brake application. When heat causes brake drum expansion, one must depress the brake pedal further to force the shoes against the drum. Industry calls this action BRAKE FADE.
Answer C is wrong because leakage past the master cylinder cup will cause a low or no pedal, not fade.
Answer D is correct because air in the hydraulic system will cause a spongy pedal, not fade.

Question #101
Answer A is wrong because soap and water causes corrosion and leaves residue.
Answer B is correct because isopropyl alcohol is the correct cleaning agent.
Answer C is wrong because mineral spirits leave residue.
Answer D is wrong because you use hydraulic brake fluid for operation of the system, not for cleaning it.

Question #102
Answer A is wrong because the proportioning valve affects only the rear wheels.
Answer B is wrong because the pressure differential valve affects the balance between the systems and activates the red warning lamp, when the pressure is different between the front and rear systems.
Answer C is correct because if the metering valve does not regulate the fluid to the front brakes, they will grab quickly when light pedal pressure is applied.
Answer D is wrong because residual pressure valve is in the master cylinder and if defective would cause a lag in braking.

Question #103
Answer A is a good choice because you do use a tool to hold the metering valve open when using a pressure bleeder to bleed the system. Yet, it is wrong because both technicians are right.
Answer B is also a good choice because you can also bleed this type of system manually. Yet, it is wrong because both technicians are right.
Answer C is correct because of the above reasons and because both technicians are right.
Answer D is wrong because both technicians are right.

Question #104
Answer A is wrong because a bell-mouth condition results when the drum diameter is greater at the edge of the drum next to the backing plate than at the edge near the wheel.
Answer B is wrong because a concave brake drum results when the drum diameter is greater in the center than at both edges.
Answer C is wrong because an out-of-round condition results when there is a variation in reading 180° apart.
Answer D is correct because a convex drum results when the diameter on a driveline parking brake drum is greater at the edges of the friction surface than in the center.

Question #105
Answer A is wrong because Federal regulations prohibit vacuum-operated power boosters on heavy trucks.
Answer B is correct because hydraulic boosters are used on heavy duty trucks.
Answer C is wrong because only one technician is right.
Answer D is wrong because one technician is right.

Question #106
Answer A is correct because the electric pump provides the power for reserve stops by recirculating fluid within the booster assembly when the engine is not running.
Answer B is wrong because no reserve electrical power is provided for the hydropower boost system.
Answer C is wrong because failure of the electric pump has no effect on the fluid circulation in the steering gear.
Answer D is wrong because braking power will not be reduced with the engine running.

Question #107
Answer A is correct because when working on air-over-hydraulic brake systems, an antifreeze valve may be used.
Answer B is wrong because a limiting valve is used on front axles to limit fluid pressure during partial braking.
Answer C is wrong because only one technician is right.
Answer D is wrong because one technician is right.

Question #108
Answer A is wrong because the intervals of repacking a truck's wheel bearings vary according to driving conditions, climate conditions, truck type, etc.
Answer B is wrong because a heavy truck's axles are of a full-floating design.
Answer C is correct because at the PM both the brakes and wheel bearings are inspected.
Answer D is wrong because wheel bearings are constructed out of steel or iron.

Question #109
Answer A is right because you torque the wheel bearing adjusting nut 50 ft.-lbs. then back off the nut one-sixth to one-third turn and install the lock ring.
Answer B is wrong because not backing off the nut will cause the bearing to overheat. You back off the nut to allow for thermal expansion.
Answer C is wrong because only one technician is correct.
Answer D is wrong because one technician is correct.

Question #110
Answer A is correct because a chassis vibration during braking can be caused by excessive radial tire runout.
Answer B is wrong because brake grab would occur on all wheels with an improper brake pedal pushrod adjustment
Answer C is wrong because brake drag is not caused by glazed brake linings. Glazed linings result in fade.
Answer D is wrong because rotors that are machined too thin will not cause a low spongy pedal.

Question #111
Answer A is correct because a swollen master cylinder cup will cause a gradual drag with complete brake lockup.
Answer B is wrong because an open compensation port does cause gradual drag with complete brake lockup. This condition causes reduced braking.
Answer C is wrong because wrong brake shoes will not cause a gradual drag with complete brake lockup.
Answer D is wrong because a weak brake shoe return spring will not cause gradual drag with complete brake lockup. The brake lockup will be more consistent.

Question #112
Answer A is wrong because you bleed the left front caliper or wheel cylinder first.
Answer B is correct because the pressure bleeder is pressurized to 20 to 25 psi (137.9–172.37 Kpa).
Answer C is wrong because the metering valve must be open.
Answer D is wrong because you bleed the left rear caliper or wheel cylinder first.

Question #113
Answer A is wrong because if the floating caliper on a hydraulic brake truck does not slide freely, excessive brake force will not result.
Answer B is wrong because excessive brake pad wear will only occur on the outboard pad.
Answer C is correct because if the floating caliper on a hydraulic brake truck does not slide freely, reduced braking force will result.
Answer D is wrong because the brakes will not grab if the floating caliper does not slide freely.

Question #114
Answer A is wrong because thickness is indicated by a code on the lining edge.
Answer B is wrong because not all semimetallic linings are edge-stamped "FF." The "FF" designation represents the cold and hot coefficient of friction.
Answer C is wrong because asbestos-type linings are no longer used.
Answer D is correct because linings should use the same edge brand as OEM.

Question #115
Answer A is correct because it is okay to machine oversize drums as long as the drum's manufacturer's recommendation for machining dimensions are followed. Yet, it is wrong because both technicians are right.
Answer B is also a good choice because turning an oversize drum will decrease its overall strength. Yet, it is wrong because both technicians are right.
Answer C is correct because both technicians are right.
Answer D is wrong because one technician is right.

Question #116
Answer A is correct because the brake application pressure should be 80 psi (553.6 Kpa).
Answer B is wrong because this test is done at every PM.
Answer C is wrong because only one technician is correct.
Answer D is wrong because one technician is correct.

Question #117
Answer A is wrong because when removing an application valve the truck should be on a level surface.
Answer B is wrong because when removing an application valve you do mark or label the brake lines.
Answer C is wrong because when removing an application valve you do mark the valve body in relation to the mounting plate.
Answer D is correct because when removing an application valve you exhaust all brake system pressure.

Question #118
Answer A is wrong because you apply air pressure to the outlet side of a single check valve.
Answer B is wrong because with air pressure applied to one of the inlet ports on a double check valve, the air pressure should be the same at the outlet port.
Answer C is correct because with air pressure applied to one inlet port in a double check valve, the test gauge at the opposite inlet port should indicate air pressure.
Answer D is wrong because when air pressure is released at one inlet port on a double check valve, the air pressure should drop quickly at the outlet port.

Question #119
Answer A is wrong because when the air pressure is reduced to zero on the delivery side of the pressure protection valve, the air pressure on the supply side should show no further pressure loss.
Answer B is wrong because the leak rate is a 1-inch bubble in 3 seconds.
Answer C is wrong because neither technician is right.
Answer D is correct because both technicians are wrong.

Question #120
Answer A is wrong because when the trailer air supply valve is pulled outward, the air pressure to the tractor protection valve is closed.
Answer B is correct because if a trailer breakaway occurs, service air pressure from the application valve can no longer pass through the tractor protection valve.
Answer C is wrong because only one technician is right.
Answer D is wrong because one technician is wrong.

Glossary

ABS An abbreviation for Anti-lock Brake System.

Absolute Pressure The aero point from which pressure is measured.

Ackerman Principle The geometric principle used to provide toe-out on turns. The ends of the steering arms are angled so that the inside wheel turns more than the outside wheel when a vehicle is making a turn.

Actuator A device that delivers motion in response to an electrical signal.

Adapter The welds under a spring seat to increase the mounting height or fit a seal to the axle.

Adapter Ring A part that is bolted between the clutch cover and the flywheel on some two-plate clutches when the clutch is installed on a flat flywheel.

A/D Converter An abbreviation for Analog-to-Digital Converter.

Additive An additive intended to improve a certain characteristic of the material.

Adjustable Torque Arm A member used to retain axle alignment and, in some cases, control axle torque. Normally one adjustable and one rigid torque arm are used per axle so the axle can be aligned. This rod has means by which it can be extended or retracted for adjustment purposes.

Adjusting Ring A device that is held in the shift signal valve bore by a press fit pin through the valve body housing. When the ring is pushed in by the adjusting tool, the slots on the ring that engage the pin are released.

After-Cooler A device that removes water and oil from the air by a cooling process. The air leaving an after-cooler is saturated with water vapor, which condenses when a drop in temperature occurs.

Air Bag An air-filled device that functions as the spring on axles that utilize air pressure in the suspension system.

Air Brakes A braking system that uses air pressure to actuate the brakes by means of diaphragms, wedges, or cams.

Air Brake System A system utilizing compressed air to activate the brakes.

Air Compressor (1) An engine-driven mechanism for supplying high pressure air to the truck brake system. There are basically two types of compressors: those designed to work on in-line engines and those that work on V-type engines. The in-line type is mounted forward and is gear driven, while the V-type is mounted toward the firewall and is camshaft driven. With both types the coolant and lubricant are supplied by the truck engine. (2) A pump-like device in the air conditioning system that compresses refrigerant vapor to achieve a change in state for the refrigeration process.

Air Conditioning The control of air movement, humidity, and temperature by mechanical means.

Air Dryer A unit that removes moisture.

Air Filter/Regulator Assembly A device that minimizes the possibility of moisture-laden air or impurities entering a system.

Air Hose An air line, such as one between the tractor and trailer, that supplies air for the trailer brakes.

Air-Over-Hydraulic Brakes A brake system utilizing a hydraulic system assisted by an air pressure system.

Air-Over-Hydraulic Intensifier A device that changes the pneumatic air pressure from the treadle brake valve into hydraulic pressure which controls the wheel cylinders.

Air Shifting The process that uses air pressure to engage different range combinations in the transmission's auxiliary section without a mechanical linkage to the driver.

Air Slide Release A release mechanism for a sliding fifth wheel, which is operated from the cab of a tractor by actuating an air control valve. When actuated, the valve energizes an air cylinder, which releases the slide lock and permits positioning of the fifth wheel.

Air Spring An airfilled device that functions as the spring on axles that utilize air pressure in the suspension system.

Air Spring Suspension A single or multi-axle suspension relying on air bags for springs and weight distribution of axles.

Air Timing The time required for the air to be transmitted to or released from each brake, starting the instant the driver moves the brake pedal.

Altitude Compensation System An altitude barometric switch and solenoid used to provide better driveability at more than 4,000 feet (1220 meters) above sea level.

Ambient Temperature Temperature of the surrounding or prevailing air. Normally, it is considered to be the temperature in the service area where testing is taking place.

Amboid Gear A gear that is similar to the hypoid type with one exception: the axis of the drive pinion gear is located above the centerline axis of the ring gear.

Amp An abbreviation for ampere.

Ampere The unit for measuring electrical current.

Analog Signal A voltage signal that varies within a given range (from high to low, including all points in between).

Analog-to-Digital Converter (A/D converter) A device that converts analog voltage signals to a digital format; this is located in a section of the processor called the input signal conditioner.

Analog Volt/Ohmmeter (AVOM) A test meter used for checking voltage and resistance. Analog meters should not be used on solid state circuits.

Annulus The largest part of a simple gear set.

Anticorrosion Agent A chemical used to protect metal surfaces from corrosion.

Antifreeze A compound, such as alcohol or glycerin, that is added to water to lower its freezing point.

Anti-lock Brake System (ABS) A computer controlled brake system having a series of sensing devices at each wheel that control braking action to prevent wheel lockup.

Anti-lock Relay Valve (ARV) In an anti-lock brake system, the device that usually replaces the standard relay valve used to control the rear axle service brakes and performs the standard relay function during tractor/trailer operation.

Antirattle Springs Springs that reduce wear between the intermediate plate and the drive pin, and helps to improve clutch release.

Antirust Agent An additive used with lubricating oils to prevent rusting of metal parts when the engine is not in use.

Application Valve A foot-operated brake valve that controls air pressure to the service chambers.

Applied Moment A term meaning a given load has been placed on a frame at a particular point.

Area The total cross section of a frame rail including all applicable elements usually given in square inches.

Armature The rotating component of a (1) starter or other motor. (2) generator. (3) compressor clutch.

Articulating Upper Coupler A bolster plate kingpin arrangement that is not rigidly attached to the trailer, but provides articulation and/or oscillation, (such as a frameless dump) about an axis parallel to the rear axle of the trailer.

Articulation Vertical movement of the front driving or rear axle relative to the frame of the vehicle to which they are attached.

ASE An abbreviation for Automotive Service Excellence, a trademark of National Institute for Automotive Service Excellence.

Aspect Ratio A tire term calculated by dividing the tire's section height by its section width.

ATEC System A system that includes an electronic control system, torque converter, lockup clutch, and planetary gear train.

Atmospheric Pressure The weight of the air at sea level; 14.696 pounds per square inch (psi) or 101.33 kilopascals (kPa).

Automatic Slack Adjuster The device that automatically adjusts the clearance between the brake linings and the brake drum or rotor. The slack adjuster controls the clearance by sensing the length of the stroke of the push rod for the air brake chamber.

Autoshift Finger The device that engages the shift blocks on the yoke bars that corresponds to the tab on the end of the gearshift lever in manual systems.

Auxiliary Filter A device installed in the oil return line between the oil cooler and the transmission to prevent debris from being flushed into the transmission causing a failure. An auxiliary filter must be installed before the vehicle is placed back in service.

Auxiliary Section The section of a transmission where range shifting occurs, housing the auxiliary drive gear, auxiliary main shaft assembly, auxiliary countershaft, and the synchronizer assembly.

Axis of Rotation The center line around which a gear or part revolves.

Axle (1) A rod or bar on which wheels turn. (2) A shaft that transmits driving torque to the wheels.

Axle Range Interlock A feature designed to prevent axle shifting when the interaxle differential is locked out, or when lockout is engaged. The basic shift system operates the same as the standard shift system to shift the axle and engage or disengage the lockout.

Axle Seat A suspension component used to support and locate the spring on an axle.

Axle Shims Thin wedges that may be installed under the leaf springs of single axle vehicles to tilt the axle and correct the U-joint operating angles. Wedges are available in a range of sizes to change pinion angles.

Backing Plate A metal plate that serves as the foundation for the brake shoes and other drum brake hardware.

Battery Terminal A tapered post or threaded studs on top of the battery case, or infernally threaded provisions on the side of the battery for connecting the cables.

Beam Solid Mount Suspension A tandem suspension relying on a pivotal mounted beam, with axles attached at the ends for load equalization. The beam is mounted to a solid center pedestal.

Beam Suspension A tandem suspension relying on a pivotally mounted beam, with axles attached at ends for lead equalization. Beam is mounted to center spring.

Bellows A movable cover or seal that is pleated or folded like an accordion to allow for expansion and contraction.

Bending Moment A term implying that when a load is applied to the frame, it will be distributed across a given section of the frame material.

Bias A tire term where belts and plies are laid diagonally or crisscrossing each other.

Bimetallic Two dissimilar metals joined together that have different bending characteristics when subjected to different changes of temperature.

Blade Fuse A type of fuse having two flat male lugs sticking out for insertion in the female box connectors.

Bleed Air Tanks The process of draining condensation from air tanks to increase air capacity and brake efficiency.

Block Diagnosis Chart A troubleshooting chart that lists symptoms, possible causes, and probable remedies in columns.

Blower Fan A fan that pushes or blows air through a ventilation, heater, or air conditioning system.

Bobtail Proportioning Valve A valve that senses when the tractor is bobtailing and automatically reduces the amount of air pressure that can be applied to the tractor's drive axle(s). This reduces braking force on the drive axles, lessening the chance of a spin out on slippery pavement.

Bobtailing A tractor running without a trailer.

Bogie The axle spring, suspension arrangement on the rear of a tandem axle tractor.

Bolster Plate The flat load-bearing surface under the front of a semitrailer, including the kingpin, which rests firmly on the fifth wheel when coupled.

Bolster Plate Height The height from the ground to the bolster plate when the trailer is level and empty.

Boss A heavy cast section that is used for support, such as the outer race of a bearing.

Bottoming A condition that occurs when; (1) The teeth of one gear touch the lowest point between teeth of a mating gear. (2) The bed or frame of the vehicle strikes the axle, such as may be the case of overloading.

Bottom U-Bolt Plate A plate that is located on the bottom side of the spring or axle and is held in place when the U-bolts are tightened to the clamp spring and axle together.

Bracket An attachment used to secure parts to the body or frame.

Brake Control Valve A dual brake valve that releases air from the service reservoirs to the service lines and brake chambers. The valve includes a piston which pushes on diaphragms to open ports; these vent air to service lines in the primary and secondary systems.

Brake Disc A steel disc used in a braking system with a caliper and pads. When the brakes are applied, the pad on each side of the spinning disc is forced against the disc, thus imparting a braking force. This type of brake is very resistant to brake fade.

Brake Drum A cast metal bell-like cylinder attached to the wheel that is used to house the brake shoes and provide a friction surface for stopping a vehicle.

Brake Fade A condition that occurs when friction surfaces become hot enough to cause the coefficient of friction to drop to a point where the application of severe pedal pressure results in little actual braking.

Brake Lining A special friction material used to line brake shoes or brake pads. It withstands high temperatures and pressure. The molded material is either riveted or bonded to the brake shoe, with a suitable coefficient of friction for stopping a vehicle.

Brake Pad The friction lining and plate assembly that is forced against the rotor to cause braking action in a disc brake system.

Brake Shoe The curved metal part, faced with brake lining, which is forced against the brake drum to produce braking action.

Brake Shoe Rollers A hardware part that attaches to the web of the brake shoes by means of roller retainers. The rollers, in turn, ride on the end of an S-cam.

Brake System The vehicle system that slows or stops a vehicle. A combination of brakes and a control system.

Breakaway Valve A device that automatically seals off the tractor air supply from the trailer air supply when the tractor system pressure drops to 30 or 40 psi (207 to 276 kPa).

British Thermal Unit (Btu) A measure of heat quantity equal to the amount of heat required to raise 1 pound of water 1°F.

Broken Back Drive Shaft A term often used for non-parallel drive shaft.

Btu An abbreviation for British Thermal Unit.

Bump Steer Erratic steering caused from rolling over bumps, cornering, or heavy braking. Same as orbital steer and roll steer.

CAA An abbreviation for Clean Air Act.

Caliper A disc brake component that changes hydraulic pressure into mechanical force and uses that force to press the brake pads against the rotor and stop the vehicle. Calipers come in three basic types: fixed, floating, and sliding, and can have one or more pistons.

Camber The attitude of a wheel and tire assembly when viewed from the front of a car. If it leans outward, away from the car at the top, the wheel is said to have positive camber. If it leans inward, it is said to have negative camber.

Cam Brakes Brakes that are similar in operation and design to the wedge brake, with the exception that an S-type camshaft is used instead of a wedge and rubber assembly.

Cartridge Fuse A type of fuse having a strip of low melting point metal enclosed in a glass tube. If an excessive current flows through the circuit, the fuse element melts at the narrow portion, opening the circuit and preventing damage.

Caster The angle formed between the kingpin axis and a vertical axis as viewed from the side of the vehicle. Caster is considered positive when the top of the kingpin axis is behind the vertical axis.

Cavitation A condition that causes bubble formation.

Center of Gravity The point around which the weight of a truck is evenly distributed; the point of balance.

Ceramic Fuse A fuse found in some import vehicles that has a ceramic insulator with a conductive metal strip along one side.

CFC An abbreviation for chlorofluorocarbon.

Charging System A system consisting of the battery, alternator, voltage regulator, associated wiring, and the electrical loads of a vehicle. The purpose of the system is to recharge the battery whenever necessary and to provide the current required to power the electrical components.

Charge the Trailer To supply the trailer air tanks with air by means of a dash control valve, tractor protection valve, and a trailer relay emergency valve.

Charging Circuit The alternator (or generator) and associated circuit used to keep the battery charged and to furnish power to the vehicle's electrical systems when the engine is running.

Check Valve A valve that allows air to flow in one direction only. It is a federal requirement to have a check valve between the wet and dry air tanks.

Chlorofluorocarbon (CFC) A compound used in the production of refrigerant that is believed to cause damage to the ozone layer.

Circuit The complete path of an electrical current, including the generating device. When the path is unbroken, the circuit is closed and current flows. When the circuit continuity is broken, the circuit is open and current flow stops.

Clean Air Act (CAA) Federal regulations, passed in 1992, that have resulted in major changes in air-conditioning systems.

Climbing A gear problem caused by excessive wear in gears, bearings, and shafts whereby the gears move sufficiently apart to cause the apex (or point) of the teeth on one gear to climb over the apex of the teeth on another gear with which it is meshed.

Clutch A device for connecting and disconnecting the engine from the transmission or for a similar purpose in other units.

Clutch Brake A circular disc with a friction surface that is mounted on the transmission input spline between the release bearing and the transmission. Its purpose is to slow or stop the transmission input shaft from rotating in order to allow gears to be engaged without clashing or grinding.

Clutch Housing A component that surrounds and protects the clutch and connects the transmission case to the vehicle's engine.

Clutch Pack An assembly of normal clutch plates, friction discs, and one very thick plate known as the pressure plate. The pressure plate has tabs around the outside diameter to mate with the channel in the clutch drum.

COE An abbreviation for cab-over-engine.

Coefficient of Friction A measurement of the amount of friction developed between two objects in physical contact when one of the objects is drawn across the other.

Coil Springs Spring steel spirals that are mounted on control arms or axles to absorb road shock.

Combination A truck coupled to one or more trailers.

Compression Applying pressure to a spring or any springy substance, thus causing it to reduce its length in the direction of the compressing force.

Compressor (1) A mechanical device that increases pressure within a container by pumping air into it. (2) That component of an air-conditioning system that compresses low temperature/pressure refrigerant vapor.

Condensation The process by which gas (or vapor) changes to a liquid.

Condenser A component in an air-conditioning system used to cool a refrigerant below its boiling point causing it to change from a vapor to a liquid.

Conductor Any material that permits the electrical current to flow.

Constant Rate Springs Leaf-type spring assemblies that have a constant rate of deflection.

Control Arm The main link between the vehicle's frame and the wheels that acts as a hinge to allow wheel action up and down independent of the chassis.

Controlled Traction A type of differential that uses a friction plate assembly to transfer drive torque from the vehicle's slipping wheel to the one wheel that has good traction or surface bite.

Converter Dolly An axle, frame, drawbar, and fifth wheel arrangement that converts a semitrailer into a full trailer.

Coolant Liquid that circulates in an engine cooling system.

Coolant Heater A component used to aid engine starting and reduce the wear caused by cold starting.

Coolant Hydrometer A tester designed to measure coolant specific gravity and determine the amount of antifreeze in the coolant.

Cooling System Complete system for circulating coolant.

Coupling Point The point at which the turbine is turning at the same speed as the impeller.

Crankcase The housing within which the crankshaft and many other parts of the engine operate.

Cranking Circuit The starter and its associated circuit, including battery, relay (solenoid), ignition switch, neutral start switch (on vehicles with automatic transmission), and cables and wires.

Cross Groove Joint disc-shaped type of inner CV joint that uses balls and V-shaped grooves on the inner and outer races to accommodate the plunging motion of the half-shaft. The joint usually bolts to a transaxle stub flange; same as disc-type joint.

Cross-Tube A system that transfers the steering motion to the opposite, passenger side steering knuckle. It links the two steering knuckles together and forces them to act in unison.

C-Train A combination of two or more trailers in which the dolly is connected to the trailer by means of two pintle hook or coupler drawbar connections. The resulting connection has one pivot point.

Cycling (1) Repeated on-off action of the air conditioner compressor. (2) Heavy and repeated electrical cycling that can cause the positive plate material to break away from its grids and fall into the sediment chambers at the base of the battery case.

Dampen To slow or reduce oscillations or movement.

Dampened Discs Discs that have dampening springs incorporated into the disc hub. When engine torque is first transmitted to the disc, the plate rotates on the hub, compressing the springs. This action absorbs the shocks and torsional vibration caused by today's low rpm, high torque, engines.

Dash Control Valves A variety of handoperated valves located on the dash. They include parking brake valves, tractor protection valves, and differential lock.

Data Links Circuits through which computers communicate with other electronic devices such as control panels, modules, some sensors, or other computers in the form of digital signals.

Dead Axle Non-live or dead axles are often mounted in lifting suspensions. They hold the axle off the road when the vehicle is traveling empty, and put it on the road when a load is being carried. They are also used as air suspension third axles on heavy straight trucks and are used extensively in eastern states with high axle weight laws. An axle that does not rotate but merely forms a base on which to attach the wheels.

Deadline To take a vehicle out of service.

Deburring To remove sharp edges from a cut.

Dedicated Contract Carriage Trucking operations set up and run according to a specific shipper's needs. In addition to transportation, they often provide other services such as warehousing and logistics planning.

Deflection Bending or moving to a new position as the result of an external force.

Department of Transportation (DOT) A government agency that establishes vehicle standards.

Detergent Additive An additive that helps keep metal surfaces clean and prevents deposits. These additives suspend particles of carbon and oxidized oil in the oil.

DER An abbreviation for Department of Environmental Resources.

Diagnostic Flow Chart A chart that provides a systematic approach to the electrical system and component troubleshooting and repair. They are found in service manuals and are vehicle make and model specific.

Dial Caliper A measuring instrument capable of taking inside, outside, depth, and step measurements.

Differential A gear assembly that transmits power from the drive shaft to the wheels and allows two opposite wheels to turn at different speeds for cornering and traction.

Differential Carrier Assembly An assembly that controls the drive axle operation.

Differential Lock A toggle or push-pull type air switch that locks together the rear axles of a tractor so they pull as one for off-the-road operation.

Digital Binary Signal A signal that has only two values; on and off.

Digital Volt/Ohmmeter (DVOM) A type of test meter recommended by most manufacturers for use on solid state circuits.

Diode The simplest semiconductor device formed by joining P-type semiconductor material with N-type semiconductor material. A diode allows current to flow in one direction, but not in the opposite direction.

Direct Drive The gearing of a transmission so that in its highest gear, one revolution of the engine produces one revolution of the transmission's output shaft. The top gear or final drive ratio of a direct drive transmission would be 1:1.

Disc Brake A steel disc used in a braking system with a caliper and pads. When the brakes are applied, the pad on each side of the spinning disc is forced against the disc, thus imparting a braking force. This type of brake is very resistant to brake fade. A type of brake that generates stopping power by the application of pads against a rotating disc (rotor).

Dispatch Sheet A form used to keep track of dates when the work is to be completed. Some dispatch sheets follow the job through each step of the servicing process.

Dog Tracking Off-center tracking of the rear wheels as related to the front wheels.

DOT An abbreviation for Department of Transportation.

Downshift Control The selection of a lower range to match driving conditions encountered or expected to be encountered. Learning to take advantage of a downshift gives better control on slick or icy roads and on steep downgrades. Downshifting to lower ranges increases engine braking.

Double Reduction Axle An axle that uses two gear sets for greater overall gear reduction and peak torque development. This design is favored for severe service applications, such as dump trucks, cement mixers, and other heavy haulers.

Drag Link A connecting rod or link between the steering gear, Pitman arm, and the steering linkage.

Drawbar Capacity The maximum, horizontal pulling force that can be safely applied to a coupling device.

Driven Gear A gear that is driven or forced to turn by a drive gear, by a shaft, or by some other device.

Drive or Driving Gear A gear that drives another gear or causes another gear to turn.

Drive Line The propeller or drive shaft, universal joints, and so forth, that links the transmission output to the axle pinion gear shaft.

Drive Line Angle The alignment of the transmission output shaft, driveshaft, and rear axle pinion centerline.

Drive Shaft An assembly of one or two universal joints connected to a shaft or tube; used to transmit power from the transmission to the differential.

Drive Train An assembly that includes all power transmitting components from the rear of the engine to the wheels, including clutch/torque converter, transmission, drive line, and front and rear driving axles.

Driver Controlled Main Differential Lock A type of axle assembly has greater flexibility over the standard type of single reduction axle because it provides equal amounts of drive line torque to each driving wheel, regardless of changing road conditions. This design also provides the necessary differential action to the road wheels when the truck is turning a corner.

Driver's Manual A publication that contains information needed by the driver to understand, operate, and care for the vehicle and its components.

Drum Brake A type of brake system in which stopping friction is created by the shoes pressing against the inside of the rotating drum.

Dual Hydraulic Braking System A brake system consisting of a tandem, or double action master cylinder which is basically two master cylinders usually formed by aligning two separate pistons and fluid reservoirs into a single cylinder.

ECU An abbreviation for electronic control unit.

Eddy Current A small circular current produced inside a metal core in the armature of a starter motor. Eddy currents produce heat and are reduced by using a laminated core.

Electricity The movement of electrons from one place to another.

Electric Retarder Electromagnets mounted in a steel frame. Energizing the retarder causes the electromagnets to exert a dragging force on the rotors in the frame and this drag force is transmitted directly to the drive shaft.

Electromotive Force (EMF) The force that moves electrons between atoms. This force is the pressure that exists between the positive and negative points (the electrical imbalance). This force is measured in units called volts.

Electronically Programmable Memory (EPROM) Computer memory that permits adaptation of the ECU to various standard mechanically controlled functions.

Electronic Control Unit (ECU) The brain of the vehicle.

Electronics The technology of controlling electricity.

Electrons Negatively charged particles orbiting around every nucleus.

Elliot Axle A solid bar front axle on which the ends span the steering knuckle.

EMF An abbreviation for electromotive force.

End Yoke The component connected to the output shaft of the transmission to transfer engine torque to the drive shaft.

Engine Brake A hydraulically operated device that converts the vehicle's engine into a power absorbing retarding mechanism.

Engine Stall Point The point, in rpms, under load is compared to the engine manufacturer's specified rpm for the stall test.

Environmental Protection Agency An agency of the United States government charged with the responsibilities of protecting the environment and enforcing the Clean Air Act (CAA) of 1990.

EPA An abbreviation for the Environmental Protection Agency.

EPROM An abbreviation for Electronically Programmable Memory.

Equalizer A suspension device used to transfer and maintain equal load distribution between two or more axles of a suspension. Formerly called a rocker beam.

Equalizer Bracket A bracket for mounting the equalizer beam of a multiple axle spring suspension to a truck or trailer frame while allowing for the beam's pivotal movement. Normally there are three basic types: flange-mount, straddle-mount, and under- or side-mount.

Evaporator A component in an air conditioning system used to remove heat from the air passing through it.

Exhaust Brake A valve in the exhaust pipe between the manifold and the muffler. A slide mechanism which restricts the exhaust flow, causing exhaust back pressure to build up in the engine's cylinders. The exhaust brake actually transforms the engine into a low pressure air compressor driven by the wheels.

External Housing Damper A counterweight attached to an arm on the rear of the transmission extension housing and designed to dampen unwanted driveline or powertrain vibrations.

Extra Capacity A term that generally refers to: (1) A coupling device that has strength capability greater than standard. (2) An oversized tank or reservoir for a fluid or vapor.

False Brinelling The polishing of a surface that is not damaged.

Fanning the Brakes Applying and releasing the brakes in rapid succession on a long downgrade.

Fatigue Failures The progressive destruction of a shaft or gear teeth material usually caused by overloading.

Fault Code A code that is recorded into the computer's memory. A fault code can be read by plugging a special break-out box tester into the computer.

Federal Motor Vehicle Safety Standard (FMVSS) A federal standard that specifies that all vehicles in the United States be assigned a Vehicle Identification Number (VIN).

Federal Motor Vehicle Safety Standard No. 121 (FMVSS 121) A federal standard that made significant changes in the guidelines that cover air brake systems. Generally speaking, the requirements of FMVSS 121 are such that larger capacity brakes and heavier steerable axles are needed to meet them.

FHWA An abbreviation for Federal Highway Administration.

Fiber Composite Springs Springs that are made of fiberglass, laminated, and bonded together by tough polyester resins.

Fifth Wheel A coupling device mounted on a truck and used to connect a semitrailer. It acts as a hinge point to allow changes in direction of travel between the tractor and the semitrailer.

Fifth Wheel Height The distance from the ground to the top of the fifth wheel when it is level and parallel with the ground. It can also refer to the height from the tractor frame to the top of the fifth wheel. The latter definition applies to data given in fifth wheel literature.

Fifth Wheel Top Plate The portion of the fifth wheel assembly that contacts the trailer bolster plate and houses the locking mechanism that connects to the kingpin.

Final Drive The last reduction gear set of a truck.

Fixed Value Resistor An electrical device that is designed to have only one resistance rating, which should not change, for controlling voltage.

Flammable Any material that will easily catch fire or explode.

Flare To spread gradually outward in a bell shape.

Flex Disc A term often used for flex plate.

Flex Plate A component used to mount the torque converter to the crankshaft. The flex plate is positioned between the engine crankshaft and the T/C. The purpose of the flex plate is to transfer crankshaft rotation to the shell of the torque converter assembly.

Float A cruising drive mode in which the throttle setting matches engine speed to road speed, neither accelerating nor decelerating.

Floating Main Shaft The main shaft consisting of a heavy-duty central shaft and several gears that turn freely when not engaged. The main shaft can move to allow for equalization of the loading on the countershafts. This is key to making a twin countershaft transmission workable. When engaged, the floating main shaft transfers torque evenly through its gears to the rest of the transmission and ultimately to the rear axle.

FMVSS An abbreviation for Federal Motor Vehicle Safety Standard.

FMVSS No 121 An abbreviation for Federal Motor Vehicle Safety Standard No 121.

Foot Valve A foot-operated brake valve that controls air pressure to the service chambers.

Foot-Pound An English unit of measurement for torque. One foot-pound is the torque obtained by a force of 1 pound applied to a foot long wrench handle.

Forged Journal Cross Part of a universal joint.

Frame Width The measurement across the outside of the frame rails of a tractor, truck, or trailer.

Franchised Dealership A dealer that has signed a contract with a particular manufacturer to sell and service a particular line of vehicles.

Fretting A result of vibration that the bearing outer race can pick up the machining pattern.

Friction Plate Assembly An assembly consisting of a multiple disc clutch that is designed to slip when a predetermined torque value is reached.

Front Axle Limiting Valve A valve that reduces pressure to the front service chambers, thus eliminating front wheel lockup on wet or icy pavements.

Front Hanger A bracket for mounting the front of the truck or trailer suspensions to the frame. Made to accommodate the end of the spring on spring suspensions. There are four basic types: flange-mount, straddle-mount, under-mount, and side-mount.

Full Trailer A trailer that does not transfer load to the towing vehicle. It employs a tow bar coupled to a swiveling or steerable running gear assembly at the front of the trailer.

Fully Floating Axles An axle configuration whereby the axle half shafts transmit only driving torque to the wheels and not bending and torsional loads that are characteristic of the semi-floating axle.

Fully Oscillating Fifth Wheel A fifth wheel type with fore/aft and side-to-side articulation.

Fusible Link A term often used for fuse link.

Fuse Link A short length of smaller gauge wire installed in a conductor, usually close to the power source.

GCW An abbreviation for gross combination weight.

Gear A disk-like wheel with external or internal teeth that serves to transmit or change motion.

Gear Pitch The number of teeth per given unit of pitch diameter, an important factor in gear design and operation.

General Over-the-Road Use A fifth wheel designed for multiple standard duty highway applications.

Gladhand The connectors between tractor and trailer air lines.

Gross Combination Weight (GCW) The total weight of a fully quipped vehicle including payload, fuel, and driver.

Gross Trailer Weight (GTW) The sum of the weight of an empty trailer and its payload.

Gross Vehicle Weight (GVW) The total weight of a fully equipped vehicle and its payload.

Ground The negatively charged side of a circuit. A ground can be a wire, the negative side of the battery, or the vehicle chassis.

Grounded Circuit A shorted circuit that causes current to return to the battery before it has reached its intended destination.

GTW An abbreviation for gross trailer weight.

GVW An abbreviation for gross vehicle weight.

Appendices

Halogen Light A lamp having a small quartz/glass bulb that contains a fuel filament surrounded by halogen gas. It is contained within a larger metal reflector and lens element.

Hand Valve (1) A valve mounted on the steering column or dash, used by the driver to apply the trailer brakes independently of the tractor brakes. (2) A hand operated valve used to control the flow of fluid or vapor.

Harness and Harness Connectors The organization of the vehicle's electrical system providing an orderly and convenient starting point for tracking and testing circuits.

Hazardous Materials Any substance that is flammable, explosive, or is known to produce adverse health effects in people or the environment that are exposed to the material during its use.

Heads Up Display (HUD) A technology used in some vehicles that superimposes data on the driver's normal field of vision. The operator can view the information, which appears to "float" just above the hood at a range near the front of a conventional tractor or truck. This allows the driver to monitor conditions such as limited road speed without interrupting his normal view of traffic.

Heater Control Valve A valve that controls the flow of coolant into the heater core from the engine.

Heat Exchanger A device used to transfer heat, such as a radiator or condenser.

Heavy-Duty Truck A truck that has a GVW of 26,001 pounds or more.

Helper Spring An additional spring device that permits greater load on an axle.

High CG Load Any application in which the load center of gravity (CG) of the trailer exceeds 40 inches (102 centimeters) above the top of the fifth wheel.

High-Resistant Circuits Circuits that have an increase in circuit resistance, with a corresponding decrease in current.

High-Strength Steel A low-alloy steel that is much stronger than hot-rolled or cold-rolled sheet steels that normally are used in the manufacture of car body frames.

Hinged Pawl Switch The simplest type of switch; one that makes or breaks the current of a single conductor.

HUD An abbreviation for heads up display.

Hydraulic Brakes Brakes that are actuated by a hydraulic system.

Hydraulic Brake System A system utilizing the properties of fluids under pressure to activate the brakes.

Hydrometer A tester designed to measure the specific gravity of a liquid.

Hypoid Gears Gears that intersect at right angles when meshed. Hypoid gearing uses a modified spiral bevel gear structure that allows several gear teeth to absorb the driving power and allows the gears to run quietly. A hypoid gear is typically found at the drive pinion gear and ring gear interface.

I-Beam Axle An axle designed to give great strength at reasonable weight. The cross section of the axle resembles the letter "I."

ICC Check Valve A valve that allows air to flow in one direction only. It is a federal requirement to have a check valve between the wet and dry air tanks.

Inboard Toward the centerline of the vehicle.

In-Line Fuse A fuse that is in series with the circuit in a small plastic fuse holder, not in the fuse box or panel. It is used, when necessary, as a protection device for a portion of the circuit even though the entire circuit may be protected by a fuse in the fuse box or panel.

In-Phase The in-line relationship between the forward coupling shaft yoke and the driveshaft slip yoke of a two-piece drive line.

Input Retarder A device located between the torque converter housing and the main housing designed primarily for over-the-road operations. The device employs a "paddle wheel" type design with a vaned rotor mounted between stator vanes in the retarder housing.

Installation Templates Drawings supplied by some vehicle manufacturers to allow the technician to correctly install the accessory. The templates available can be used to check clearances or to ease installation.

Insulator A material, such as rubber or glass, that offers high resistance to the flow of electrons.

Integrated Circuit A component containing diodes, transistors, resistors, capacitors, and other electronic components mounted on a single piece of material and capable to perform numerous functions.

Jacobs Engine Brake A term sometimes used for Jake brake.

Jake Brake The Jacobs engine brake, named for its inventor. A hydraulically operated device that converts a power producing diesel engine into a power-absorbing retarder mechanism by altering the engine's exhaust valve opening time used to slow the vehicle.

Jumper Wire A wire used to temporarily bypass a circuit or components for electrical testing. A jumper wire consists of a length of wire with an alligator clip at each end.

Jumpout A condition that occurs when a fully engaged gear and sliding clutch are forced out of engagement.

Jump Start The procedure used when it becomes necessary to use a booster battery to start a vehicle having a discharged battery.

Kinetic Energy Energy in motion.

Kingpin (1) The pin mounted through the center of the trailer upper coupler (bolster plate) that mates with the fifth wheel locks, securing the trailer to the fifth wheel. The configuration is controlled by industry standards. (2) A pin or shaft on which the steering spindle rotates.

Landing Gear The retractable supports for a semitrailer to keep the trailer level when the tractor is detached from it.

Lateral Runout The wobble or side-to-side movement of a rotating wheel or of a rotating wheel and tire assembly.

Lazer Beam Alignment System A two- or four-wheel alignment system using wheel-mounted instruments to project a lazer beam to measure toe, caster, and camber.

Lead The tendency of a car to deviate from a straight path on a level road when there is no pressure on the steering wheel in either direction.

Leaf Springs Strips of steel connected to the chassis and axle to isolate the vehicle from road shock.

Less Than Truckload (LTL) Partial loads from the networks of consolidation centers and satellite terminals.

Light Beam Alignment System An alignment system using wheel-mounted instruments to project light beams onto charts and scales to measure toe, caster, and camber, and note the results of alignment adjustments.

Limited-Slip Differential A differential that utilizes a clutch device to deliver power to either rear wheel when the opposite wheel is spinning.

Linkage A system of rods and levers used to transmit motion or force.

Live Axle An axle on which the wheels are firmly affixed. The axle drives the wheels.

Live Beam Axle A non-independent suspension in which the axle moves with the wheels.

Load Proportioning Valve (LPV) A valve used to redistribute hydraulic pressure to front and rear brakes based on vehicle loads. This is a load- or height-sensing valve that senses the vehicle load and proportions the braking between front and rear brakes in proportion to the load variations and degree of rear-to-front weight transfer during braking.

Lockstrap A manual adjustment mechanism that allows for the adjustment of free travel.

Lock-up Torque Converter A torque converter that eliminates the 10 percent slip that takes place between the impeller and turbine at the coupling stage of operation. It is considered a four-element (impeller, turbine, stator, lockup clutch), three-stage (stall, coupling, and locking stage) unit.

Longitudinal Leaf Spring A leaf spring that is mounted so it is parallel to the length of the vehicle.

Low-Maintenance Battery A conventionally vented, lead/acid battery, requiring normal periodic maintenance.

LTL An abbreviation for less than truckload.

Magnetorque An electromagnetic clutch.

Maintenance-Free Battery A battery that does not require the addition of water during normal service life.

Maintenance Manual A publication containing routine maintenance procedures and intervals for vehicle components and systems.

Main Transmission A transmission consisting of an input shaft, floating main shaft assembly and main drive gears, two counter shaft assemblies, and reverse idler gears.

Manual Slide Release The release mechanism for a sliding fifth wheel, which is operated by hand.

Metering Valve A valve used on vehicles equipped with front disc and rear drum brakes. It improves braking balance during light brake applications by preventing application of the front disc brakes until pressure is built up in the hydraulic system.

Moisture Ejector A valve mounted to the bottom or side of the supply and service reservoirs that collects water and expels it every time the air pressure fluctuates.

Mounting Bracket That portion of the fifth wheel assembly that connects the fifth wheel top plate to the tractor frame or fifth wheel mounting system.

Multiaxle Suspension A suspension consisting of more than three axles.

Multiple Disc Clutch A clutch having a large drum-shaped housing that can be either a separate casting or part of the existing transmission housing.

NATEF An abbreviation for National Automotive Education Foundation.

National Automotive Education Foundation (NATEF) A foundation having a program of certifying secondary and post secondary automotive and heavy-duty truck training programs.

National Institute for Automotive Service Excellence (ASE) A nonprofit organization that has an established certification program for automotive, heavy-duty truck, auto body repair, engine machine shop technicians, and parts specialists.

Needlenose Pliers This tool has long tapered jaws for grasping small parts or for reaching into tight spots. Many needlenose pliers also have cutting edges and a wire stripper.

NIASE An abbreviation for National Institute for Automotive Service Excellence, now abbreviated ASE.

NIOSH An abbreviation for National Institute for Occupation Safety and Health.

NLGI An abbreviation for National Lubricating Grease Institute.

NHTSA An abbreviation for National Highway Traffic Safety Administration.

Nonlive Axle Non-live or dead axles are often mounted in lifting suspensions. They hold the axle off the road when the vehicle is traveling empty, and put it on the road when a load is being carried. They are also used as air suspension third axles on heavy straight trucks and are used extensively in eastern states with high axle weight laws.

Nonparallel Driveshaft A type of drive shaft installation whereby the working angles of the joints of a given shaft are equal; however the companion flanges and/or yokes are not parallel.

Nonpolarized Gladhand A gladhand that can be connected to either service or emergency gladhand.

Nose The front of a semitrailer.

No-tilt Convertible Fifth A fifth wheel with fore/aft articulation that can be locked out to produce a rigid top plate for applications that have either rigid and/or articulating upper couplers.

(OEM) An abbreviation for original equipment manufacturer.

Off-road With reference to unpaved, rough, or ungraded terrain on which a vehicle will operate. Any terrain not considered part of the highway system falls into this category.

Ohm A unit of measured electrical resistance.

Ohm's Law The basic law of electricity stating that in any electrical circuit, current, resistance, and pressure work together in a mathematical relationship.

On-road With reference to paved or smooth-graded surface terrain on which a vehicle will operate, generally considered to be part of the public highway system.

Open Circuit An electrical circuit whose path has been interrupted or broken either accidentally (a broken wire) or intentionally (a switch turned off).

Operational Control Valve A valve used to control the flow of compressed air through the brake system.

Oscillation The rotational movement in either fore/aft or side-to-side direction about a pivot point. Generally refers to fifth wheel designs in which fore/aft and side-to-side articulation are provided.

OSHA An abbreviation for Occupational Safety and Health Administration.

Out-of-Phase A condition of the universal joint which acts somewhat like one person snapping a rope held by a person at the opposite end. The result is a violent reaction at the opposite end. If both were to snap the rope at the same time, the resulting waves cancel each other and neither would feel the reaction.

Out-of-Round A wheel or tire defect in which the wheel or tire is not round.

Output Driver An electronic on/off switch that the computer uses to control the ground circuit of a specific actuator. Output drivers are located in the processor along with the input conditioners, microprocessor, and memory.

Output Yoke The component that serves as a connecting link, transferring torque from the transmission's output shaft through the vehicle's drive line to the rear axle.

Oval A condition that occurs when a tube is not round, but is somewhat egg-shaped.

Overall Ratio The ratio of the lowest to the highest forward gear in the transmission.

Overdrive The gearing of a transmission so that in its highest gear one revolution of the engine produces more than one revolution of the transmission's output shaft.

Overrunning Clutch A clutch mechanism that transmits power in one direction only.

Overspeed Governor A governor that shuts off the fuel or stops the engine when excessive speed is reached.

Oxidation Inhibitor (1) An additive used with lubricating oils to keep oil from oxidizing even at very high temperatures. (2) An additive for gasoline to reduce the chemicals in gasoline that react with oxygen.

Pad A disc brake lining and metal back riveted, molded, or bonded together.

Parallel Circuit An electrical circuit that provides two or more paths for the current to flow. Each path has separate resistors and operates independently from the other parallel paths. In a parallel circuit, amperage can flow through more than one resistor at a time.

Parallel Joint Type A type of drive shaft installation whereby all companion flanges and/or yokes in the complete drive line are parallel to each other with the working angles of the joints of a given shaft being equal and opposite.

Parking Brake A mechanically applied brake used to prevent a parked vehicle's movement.

Parts Requisition A form that is used to order new parts, on which the technician writes the names of what part(s) are needed along with the vehicle's VIN or company's identification folder.

Payload The weight of the cargo carried by a truck, not including the weight of the body.

Pipe or Angle Brace Extrusions between opposite hangers on a spring or air-type suspension.

Pitman Arm A steering linkage component that connects the steering gear to the linkage at the left end of the center link.

Pitting Surface irregularities resulting from corrosion.

Planetary Drive A planetary gear reduction set where the sun gear is the drive and the planetary carrier is the output.

Planetary Gear Set A system of gearing that is somewhat like the solar system. A pinion is surrounded by an internal ring gear and planet gears are in mesh between the ring gear and pinion around which all revolve.

Planetary Pinion Gears Small gears fitted into a framework called the planetary carrier.

Plies The layers of rubber-impregnated fabric that make up the body of a tire.

Pogo Stick The air and electrical line support rod mounted behind the cab to keep the lines from dragging between the tractor and trailer.

Polarity The particular state, either positive or negative, with reference to the two poles or to electrification.

Pole The number of input circuits made by an electrical switch.

Pounds per Square Inch (psi) A unit of English measure for pressure.

Power A measure of work being done.

Power Flow The flow of power from the input shaft through one or more sets of gears, or through an automatic transmission to the output shaft.

Power Steering A steering system utilizing hydraulic pressure to reduce the turning effort required of the operator.

Power Synchronizer A device to speed up the rotation of the main section gearing for smoother automatic downshifts and to slow down the rotation of the main section gearing for smoother automatic upshifts.

Power Train An assembly consisting of a drive shaft, coupling, clutch, and transmission differential.

Pressure The amount of force applied to a definite area measured in pounds per square inch (psi) English or kilopascals (kPa) metric.

Pressure Differential The difference in pressure between any two points of a system or a component.

Pressure Relief Valve (1) A valve located on the wet tank, usually preset at 150 psi (1,034 kPa). Limits system pressure if the compressor or governor unloader valve malfunctions. (2) A valve located on the rear head of an air-conditioning compressor or pressure vessel that opens if an excessive system pressure is exceeded.

Printed Circuit Board An electronic circuit board made of thin nonconductive plastic-like material onto which conductive metal, such as copper, has been deposited. Parts of the metal are then etched away by an acid, leaving metal lines that form the conductors for the various circuits on the board. A printed circuit board can hold many complex circuits in a very small area.

Programmable Read Only Memory (PROM) An electronic component that contains program information specific to different vehicle model calibrations.

PROM An abbreviation for Programmable Read Only Memory.

Priority Valve A valve that ensures that the control system upstream from the valve will have sufficient pressure during shifts to perform its automatic functions.

Proportioning Valve A valve used on vehicles equipped with front disc and rear drum brakes. It is installed in the lines to the rear drum brakes, and in a split system, below the pressure differential valve. By reducing pressure to the rear drum brakes, the valve helps to prevent premature lockup during severe brake application and provides better braking balance.

Psi An abbreviation for pounds per square inch.

Pull Circuit A circuit that brings the cab from a fully tilted position up and over the center.

Pull-Type Clutch A type of clutch that does not push the release bearing toward the engine; instead, it pulls the release bearing toward the transmission.

Pump/Impeller Assembly The input (drive) member that receives power from the engine.

Push Circuit A circuit that raises the cab from the lowered position to the desired tilt position.

Push-Type Clutch A type of clutch in which the release bearing is not attached to the clutch cover.

P-type Semiconductors Positively charged materials that enables them to carry current. They are produced by adding an impurity with three electrons in the outer ring (trivalent atoms).

Quick Release Valve A device used to exhaust air as close as possible to the service chambers or spring brakes.

Radial A tire design having cord materials running in a direction from the center point of the tire, usually from bead to bead.

Radial Load A load that is applied at 90° to an axis of rotation.

RAM An abbreviation for random access memory.

Ram Air Air that is forced into the engine or passenger compartment by the forward motion of the vehicle.

Random Access Memory (RAM) The memory used during computer operation to store temporary information. The microcomputer can write, read, and erase information from RAM in any order, which is why it is called random.

Range Shift Cylinder A component located in the auxiliary section of the transmission. This component, when directed by air pressure via low and high ports, shifts between high and low range of gears.

Range Shift Lever A lever located on the shift knob allows the driver to select low or high gear range.

Rated Capacity The maximum, recommended safe load that can be sustained by a component or an assembly without permanent damage.

Ratio Valve A valve used on the front or steering axle of a heavy-duty truck to limit the brake application pressure to the actuators during normal service braking.

RCRA An abbreviation for Resource Conservation and Recovery Act.

Reactivity The characteristic of a material that enables it to react violently with air, heat, water, or other materials.

Read Only Memory (ROM) A type of memory used in microcomputers to store information permanently.

Rear Hanger A bracket for mounting the rear of a truck or trailer suspension to the frame. Made to accommodate the end of the spring on spring suspensions. There are usually four types: flange-mount, straddle-mount, under-mount, and side-mount.

Recall Bulletin A bulletin that pertains to special situations that involve service work or replacement of parts in connection with a recall notice.

Reference Voltage The voltage supplied to a sensor by the computer, which acts as a base line voltage; modified by the sensor to act as an input signal.

Relay An electric switch that allows a small current to control a much larger one. It consists of a control circuit and a power circuit.

Relay/Quick Release Valve A valve used on trucks with a wheel base 254 inches (6.45 meters) or longer. It is attached to an air tank to main supply line to speed the application and release of air to the service chambers. It is similar to a remote control foot valve.

Refrigerant A liquid capable of vaporizing at a low temperature.

Refrigerant Management Center Equipment designed to recover, recycle, and recharge an air-conditioning system.

Release Bearing A unit within the clutch consisting of bearings that mount on the transmission input shaft but do not rotate with it.

Reserve Capacity Rating The ability of a battery to sustain a minimum vehicle electrical load in the event of a charging system failure.

Resistance The opposition to current flow in an electrical circuit.

Resisting Bending Moment A measurement of frame rail strength derived by multiplying the section modulus of the rail by the yield strength of the material. This term is universally used in evaluating frame rail strength.

Resource Conservation and Recovery Act (RCRA) A law that states that after using a hazardous material, it must be properly stored until an approved hazardous waste hauler arrives to take them to the disposal site.

Reverse Elliot Axle A solid-beam front axle on which the steering knuckles span the axle ends.

Revolutions per Minute (rpm) The number of complete turns a member makes in one minute.

Right to Know Law A law passed by the federal government and administered by the Occupational Safety and Health Administration (OSHA) that requires any company that uses or produces hazardous chemicals or substances to inform its employees, customers, and vendors of any potential hazards that may exist in the workplace as a result of using the products.

Rigid Disc A steel plate to which friction linings, or facings, are bonded or riveted.

Rigid Fifth Wheel A platform that is fixed rigidly to a frame. This fifth wheel has no articulation or oscillation. It is generally used in applications where the articulation is provided by other means, such as an articulating upper coupler of a frame-less dump.

Rigid Torque Arm A member used to retain axle alignment and, in some cases, to control axle torque. Normally, one adjustable and one rigid arm are used per axle so the axle can be aligned.

Ring Gear (1) The gear around the edge of a flywheel. (2) A large circular gear such as that found in a final drive assembly.

Rocker Beam A suspension device used to transfer and maintain equal load distribution between two or more axles of a suspension.

Roll Axis The theoretical line that joins the roll center of the front and rear axles.

Roller Clutch A clutch designed with a movable inner race, rollers, accordion (apply) springs, and outer race. Around the inside diameter of the outer race are several cam-shaped pockets. The clutch assembly rollers and accordion springs are located in these pockets.

Rollers A hardware part that attaches to the web of the brake shoes by means of roller retainers. The rollers, in turn, ride on the end of an S-cam.

ROM An abbreviation for read only memory.

Rotary Oil Flow A condition caused by the centrifugal force applied to the fluid as the converter rotates around its axis.

Rotation A term used to describe the fact that a gear, shaft, or other device is turning.

rpm An abbreviation for revolutions per minute.

Rotor (1) A part of the alternator that provides the magnetic fields necessary to create a current flow. (2) The rotating member of an assembly.

Runout A deviation of the specified normal travel of an object. The amount of deviation or wobble a shaft or wheel has as it rotates. Runout is measured with a dial indicator.

Safety Factor (SF) (1) The amount of load which can safely be absorbed by and through the vehicle chassis frame members. (2) The difference between the stated and rated limits of a product, such as a grinding disk.

Screw Pitch Gauge A gauge used to provide a quick and accurate method of checking the threads per inch of a nut or bolt.

Secondary Lock The component or components of a fifth wheel locking mechanism that can be included as a backup system for the primary locks. The secondary lock is not required for the fifth wheel to function and can be either manually or automatically applied. On some designs, the engagement of the secondary lock can only be accomplished if the primary lock is properly engaged.

Section Height The tread center to bead plane on a tire.

Section Width The measurement on a tire from sidewall to sidewall.

Self-Adjusting Clutch A clutch that automatically takes up the slack between the pressure plate and clutch disc as wear occurs.

Semiconductor A solid state device that can function as either a conductor or an insulator, depending on how its structure is arranged.

Semifloating Axle An axle type whereby drive power from the differential is taken by each axle half-shaft and transferred directly to the wheels. A single bearing assembly, located at the outer end of the axle, is used to support the axle half-shaft.

Semioscillating A term that generally describes a fifth wheel type that oscillates or articulates about an axis perpendicular to the vehicle centerline.

Semitrailer A load-carrying vehicle equipped with one or more axles and constructed so that its front end is supported on the fifth wheel of the truck tractor that pulls it.

Sensing Voltage The voltage that allows the regulator to sense and monitor the battery voltage level.

Sensor An electronic device used to monitor relative conditions for computer control requirements.

Series Circuit A circuit that consists of two or more resistors connected to a voltage source with only one path for the electrons to follow.

Series/Parallel Circuit A circuit designed so that both series and parallel combinations exist within the same circuit.

Service Bulletin A publication that provides the latest service tips, field repairs, product improvements, and related information of benefit to service personnel.

Service Manual A manual, published by the manufacturer, that contains service and repair information for all vehicle systems and components.

Shift Bar Housing Available in standard- and forward-position configurations, a component that houses the shift rails, shift yokes, detent balls and springs, inter-lock balls, and pin and neutral shaft.

Shift Fork The Y-shaped component located between the gears on the main shaft that, when actuated, cause the gears to engage or disengage via the sliding clutches. Shift forks are located between low and reverse, first and second, and third and fourth gears.

Shift Rail Shift rails guide the shift forks using a series of grooves, tension balls, and springs to hold the shift forks in gear. The grooves in the forks allow them to interlock the rails, and the transmission cannot be accidentally shifted into two gears at the same time.

Shift Tower The main interface between the driver and the transmission, consisting of a gearshift lever, pivot pin, spring, boot and housing.

Shift Yoke A Y-shaped component located between the gears on the main shaft that, when actuated, cause the gears to engage or disengage via the sliding clutches. Shift yokes are located between low and reverse, first and second, and third and fourth gears.

Shock Absorber A hydraulic device used to dampen vehicle spring oscillations for controlling body sway and wheel bounce, and/or prevent spring breakage.

Short Circuit An undesirable connection between two worn or damaged wires. The short occurs when the insulation is worn between two adjacent wires and the metal in each wire contacts the other, or when the wires are damaged or pinched.

Single-Axle Suspension A suspension with one axle.

Single Reduction Axle Any axle assembly that employs only one gear reduction through its differential carrier assembly.

Slave Valve A valve to help protect gears and components in the transmission's auxiliary section by permitting range shifts to occur only when the transmission's main gearbox is in neutral. Air pressure from a regulator signals the slave valve into operation.

Slide Travel The distance that a sliding fifth wheel is designed to move.

Sliding Fifth Wheel A specialized fifth wheel design that incorporates provisions to readily relocate the kingpin center forward and rearward, which affects the weight distribution on the tractor axles and/or overall length of the tractor and trailer.

Slipout A condition that generally occurs when pulling with full power or decelerating with the load pushing. Tapered or worn clutching teeth will try to "walk" apart as the gears rotate, causing the sliding clutch and gear to slip out of engagement.

Slip Rings and Brushes Components of an alternator that conducts current to the rotor. Most alternators have two slip rings mounted directly on the rotor shaft; they are insulated from the shaft and from each other. A spring loaded carbon brush is located on each slip ring to carry the current to and from the rotor windings.

Solenoid An electromagnet that is used to perform work, made with one or two coil windings wound around an iron tube.

Solid-State Device A device that requires very little power to operate, is very reliable, and generates very little heat.

Solid Wires A single-strand conductor.

Solvent A substance which dissolves other substances.

Spade Fuse A term used for blade fuse.

Spalling Surface fatigue occurs when chips, scales, or flakes of metal break off due to fatigue rather than wear. Spalling is usually found on splines and U-joint bearings.

Specialty Service Shop A shop that specializes in areas such as engine rebuilding, transmission/axle overhauling, brake, air conditioning/heating repairs, and electrical/electronic work.

Specific Gravity The scientific measurement of a liquid based on the ratio of the liquid's mass to an equal volume of distilled water.

Spiral Bevel Gear A gear arrangement that has a drive pinion gear that meshes with the ring gear at the centerline axis of the ring gear. This gearing provides strength and allows for quiet operation.

Splined Yoke A yoke that allows the drive shaft to increase in length to accommodate movements of the drive axles.

Spontaneous Combustion A process by which a combustible material ignites by itself and starts a fire.

Spread Tandem Suspension A two-axle assembly in which the axles are spaced to allow maximum axle loads under existing regulations. The distance is usually more than 55 inches.

Spring A device used to reduce road shocks and transfer loads through suspension components to the frame of the trailer. There are usually four basic types: multileaf, monoleaf, taper, and air springs.

Spring Chair A suspension component used to support and locate the spring on an axle.

Spring Deflection The depression of a trailer suspension when the springs are placed under load.

Spring Rate The load required to deflect the spring a given distance, (usually one inch).

Spring Spacer A riser block often used on top of the spring seat to obtain increased mounting height.

Stability A relative measure of the handling characteristics which provide the desired and safe operation of the vehicle during various maneuvers.

Stabilizer A device used to stabilize a vehicle during turns; sometimes referred to as a sway bar.

Stabilizer Bar A bar that connects the two sides of a suspension so that cornering forces on one wheel are shaped by the other. This helps equalize wheel side loading and reduces the tendency of the vehicle body to roll outward in a turn.

Staff Test A test performed when there is an obvious malfunction in the vehicle's power package (engine and transmission), to determine which of the components is at fault.

Stand Pipe A type of check valve which prevents reverse flow of the hot liquid lubricant generated during operation. When the universal joint is at rest, one or more of the cross ends will be up. Without the stand pipe, lubricant would flow out of the upper passage ways and trunnions, leading to partially dry startup.

Starter Circuit The circuit that carries the high current flow within the system and supplies power for the actual engine cranking.

Starter Motor The device that converts the electrical energy from the battery into mechanical energy for cranking the engine.

Starting Safety Switch A switch that prevents vehicles with automatic transmissions from being started in gear.

Static Balance Balance at rest, or still balance. It is the equal distribution of the weight of the wheel and tire around the axis of rotation so that the wheel assembly has no tendency to rotate by itself regardless of its position.

Stationary Fifth Wheel A fifth wheel whose location on the tractor frame is fixed once it is installed.

Stator A component located between the pump/impeller and turbine to redirect the oil flow from the turbine back into the impeller in the direction of impeller rotation with minimal loss of speed or force.

Stator Assembly The reaction member or torque multiplier supported on a free wheel roller race that is splined to the valve and front support assembly.

Steering Gear A gear set mounted in a housing that is fastened to the lower end of the steering column used to multiply driver turning force and change rotary motion into longitudinal motion.

Steering Stabilizer A shock absorber attached to the steering components to cushion road shock in the steering system, improving driver control in rough terrain and protecting the system.

Stepped Resistor A resistor designed to have two or more fixed values, available by connecting wires to either of the several taps.

Still Balance Balance at rest; the equal distribution of the weight of the wheel and tire around the axis of rotation so that the wheel assembly has no tendency to rotate by itself regardless of its position.

Stoplight Switch A pneumatic switch that actuates the brake lights. There are two types: (1) A service stoplight switch that is located in the service circuit, actuated when the service brakes are applied. (2) An emergency stoplight switch located in the emergency circuit and actuated when a pressure loss occurs.

Storage Battery A battery to provide a source of direct current electricity for both the electrical and electronic systems.

Stranded Wire Wire that is are made up of a number of small solid wires, generally twisted together, to form a single conductor.

Structural Member A primary load-bearing portion of the body structure that affects its over-the-road performance or crash-worthiness.

Sulfation A condition that occurs when sulfate is allowed to remain in the battery plates for a long time, causing two problems: (1) It lowers the specific gravity levels, increasing the danger of freezing at low temperatures. (2) In cold weather a sulfated battery may not have the reserve power needed to crank the engine.

Suspension A system whereby the axle or axles of a unit are attached to the vehicle frame, designed in such a manner that road shocks from the axles are dampened through springs reducing the forces entering the frame.

Suspension Height The distance from a specified point on a vehicle to the road surface when not at curb weight.

Swage To reduce or taper.

Sway Bar A component that connects the two sides of a suspension so that cornering forces on one wheel are shared by the other. This helps equalize wheel side loading and reduces the tendency of the vehicle body to roll outward in a turn.

Switch A device used to control on/off and direct the flow of current in a circuit. A switch can be under the control of the driver or can be self-operating through a condition of the circuit, the vehicle, or the environment.

Synchromesh A mechanism that equalizes the speed of the gears that are clutched together.

Appendices — Glossary

Synchro-transmission A transmission with mechanisms for synchronizing the gear speeds so that the gears can be shifted without clashing, thus eliminating the need for double-clutching.

System Protection Valve A valve to protect the brake system against an accidental loss of air pressure, buildup of excess pressure, or back-flow and reverse air flow.

Tachometer An instrument that indicates rotating speeds, sometimes used to indicate crankshaft rpm.

Tag Axle The rearmost axle of a tandem axle tractor used to increase the load-carrying capacity of the vehicle.

Tapped Resistor A resistor designed to have two or more fixed values, available by connecting wires to either of the several taps.

Tandem One directly in front of the other and working together.

Tandem Axle Suspension A suspension system consisting of two axles with a means for equalizing weight between them.

Tandem Drive A two-axle drive combination.

Tandem Drive Axle A type of axle that combines two single axle assemblies through the use of an interaxle differential or power divider and a short shaft that connects the two axles together.

Three-Speed Differential A type of axle in a tandem two-speed axle arrangement with the capability of operating the two drive axles in different speed ranges at the same time. The third speed is actually an intermediate speed between the high and low range.

Throw (1) The offset of a crankshaft. (2) The number of output circuits of a switch.

Tie-Rod Assembly A system that transfers the steering motion to the opposite, passenger side steering knuckle. It links the two steering knuckles together and forces them to act in unison.

Time Guide Prepared reference material used for computing compensation payable by the truck manufacturer for repairs or service work to vehicles under warranty, or for other special conditions authorized by the company.

Timing (1) A procedure of marking the appropriate teeth of a gear set prior to installation and placing them in proper mesh while in the transmission. (2) The combustion spark delivery in relation to the piston position.

Toe A suspension dimension that reflects the difference in the distance between the extreme front and rear of the tire.

Toe In A suspension dimension whereby the front of the tire points inward toward the vehicle.

Toe Out A suspension dimension whereby the front of the tire points outward from the vehicle.

Top U-Bolt Plate A plate located on the top of the spring and is held in place when the U-bolts are tightened to clamp the spring and axle together.

Torque To tighten a fastener to a specific degree of tightness, generally in a given order or pattern if multiple fasteners are involved on a single component.

Torque and Twist A term that generally refers to the forces developed in the trailer and/or tractor frame that are transmitted through the fifth wheel when a rigid trailer, such as a tanker, is required to negotiate bumps, like street curbs.

Torque Converter A component device, similar to a fluid coupling, that transfers engine torque to the transmission input shaft and can multiply engine torque by having one or more stators between the members.

Torque Limiting Clutch Brake A system designed to slip when loads of 20 to 25 pound–feet (27 to 34 N) are reached protecting the brake from overloading and the resulting high heat damage.

Torque Rod Shim A thin wedge-like insert that rotates the axle pinion to change the U-joint operating angle.

Torsional Rigidity A component's ability to remain rigid when subjected to twisting forces.

Torsion Bar Suspension A type of suspension system that utilizes torsion bars in lieu of steel leaf springs or coil springs. The typical torsion bar suspension consists of a torsion bar, front crank, and rear crank with associated brackets, a shackle pin, and assorted bushings and seals.

Total Pedal Travel The complete distance the clutch or brake pedal must move.

Toxicity A statement of how poisonous a substance is.

Tracking The travel of the rear wheels in a parallel path with the front wheels.

Tractor A motor vehicle, without a body, that has a fifth wheel and is used for pulling a semitrailer.

Tractor Protection Valve A device that automatically seals off the tractor air supply from the trailer air supply when the tractor system pressure drops to 30 or 40 psi (207 to 276 kPa).

Tractor/Trailer Lift Suspension A single axle air ride suspension with lift capabilities commonly used with steerable axles for pusher and tag applications.

Trailer A platform or container on wheels pulled by a car, truck, or tractor.

Trailer Hand Control Valve A device located on the dash or steering column and used to apply only the trailer brakes; primarily used in jackknife situations.

Trailer Slider A movable trailer suspension frame that is capable of changing trailer wheelbase by sliding and locking into different positions.

Transfer Case An additional gearbox located between the main transmission and the rear axle to transfer power from the transmission to the front and rear driving axles.

Transistor An electronic device produced by joining three sections of semiconductor materials. Like the diode, it is very useful as a switching device, functioning as either a conductor or an insulator.

Transmission A device used to transmit torque at various ratios and that can usually also change the direction of the force of rotation.

Transverse Vibrations A condition caused by an unbalanced driveline or bending movements, in the drive shaft.

Treadle A dual brake valve that releases air from the service reservoirs to the service lines and brake chambers. The valve includes a piston which pushes on diaphragms to open ports; these vent air to service lines in the primary and secondary systems.

Treadle Valve A foot-operated brake valve that controls air pressure to the service chambers.

Tree Diagnosis Chart A chart used to provide a logical sequence for what should be inspected or tested when troubleshooting a repair problem.

Triaxle Suspension A suspension consisting of three axles with a means of equalizing weight between axles.

Trunnion The end of the universal cross; they are case hardened ground surfaces on which the needle bearings ride.

TTMA An abbreviation for Truck and Trailer Manufacturers Association.

Turbine The output (driven) member that is splined to the forward clutch of the transmission and to the turbine shaft assembly.

TVW An abbreviation for (1) Total vehicle weight. (2) Towed vehicle weight.

Two-Speed Axle Assembly An axle assembly having two different output ratios from the differential. The driver selects the ratios from the controls located in the cab of the truck.

U-Bolt A fastener used to clamp the top U-bolt plate, spring, axle, and bottom U-bolt plate together. Inverted (nuts down) U-bolts cross springs when in place; conventional (nuts up) U-bolts wrap around the axle.

UNEP An abbreviation for United Nations Environment Program. Mandates the complete phaseout of CFC-based refrigerants by 1995.

Underslung Suspension A suspension in which the spring is positioned under the axle.

United Nations Environmental Program (UNEP) A protocol that mandated the complete phase-out of CFC-based refrigerants by the year 1995.

Universal Gladhand A term often used for non-polarized gladhand.

Universal Joint (U-joint) A component that allows torque to be transmitted to components that are operating at different angles.

Upper Coupler The flat load-bearing surface under the front of a semitrailer, including the kingpin, which rests firmly on the fifth wheel when coupled.

Vacuum Air below atmospheric pressure. There are three types of vacuums important to engine and component function: manifold vacuum, ported vacuum, and venturi vacuum. The strength of either of these vacuums depend on throttle opening, engine speed, and load.

Validity List A list supplied by the manufacturer of valid bulletins.

Valve Body and Governor Test Stand A specialized piece of test equipment. The valve body of the transmission is removed from the vehicle and mounted into the test stand. The test stand duplicates all vehicle running conditions, so the valve body can be thoroughly tested and calibrated.

Variable Pitch Stator A stator design often used in torque converters in off-highway applications such as dirt and stone aggregate dump or haul trucks, or other specialized equipment used to transport unusually heavy loads in rough terrain.

Vehicle Body Clearance (VBC) The distance from the inside of the inner tire to the spring or other body structures.

Vehicle On-board Radar (VORAD) A system similar to an electronic eye that constantly monitors other vehicles on the road to give the driver additional reaction time to respond to potential dangers.

Vehicle Retarder An optional type of braking device that has been developed and successfully used over the years to supplement or assist the service brakes on heavy-duty trucks.

Vertical Load Capacity The maximum, recommended vertical downward force that can be safely applied to a coupling device.

VIN An abbreviation for Vehicle Identification Number.

Viscosity The ability of an oil to maintain proper lubricating quality under various conditions of operating speed, temperature, and pressure. Viscosity describes oil thickness or resistance to flow.

Volt The unit of electromotive force.

Voltage Generating Sensors These are devices which produces their own input voltage signal.

Voltage Limiter A device that provides protection by limiting voltage to the instrument panel gauges to approximately 5 volts.

Voltage Regulator A device that controls the amount of current produced by the alternator or generator and thus the voltage level in the charging circuit.

VORAD An acronym for Vehicle On-board Radar.

Vortex Oil Flow The circular flow that occurs as the oil is forced from the impeller to the turbine and then back to the impeller.

Watt The measure of electrical power.

Watt's Law A basic law of electricity used to find the power of an electrical circuit expressed in watts. It states that power equals the voltage multiplied by the current, in amperes.

Wear Compensator A device mounted in the clutch cover having an actuator arm that fits into a hole in the release sleeve retainer.

Wedge-Actuated Brakes A brake system using air pressure and air brake chambers to push a wedge and roller assembly into an actuator that is located between adjusting and anchor pistons.

Wet Tank A supply reservoir.

Wheel Alignment The mechanics of keeping all the parts of the steering system in the specified relation to each other.

Wheel and Axle Speed Sensors Electromagnetic devices used to monitor vehicle speed information for an antilock controller.

Wheel Balance The equal distribution of weight in a wheel with the tire mounted. It is an important factor which affects tire wear and vehicle control.

Windings (1) The three separate bundles in which wires are grouped in the stator. (2) The coil of wire found in a relay or other similar device. (3) That part of an electrical clutch that provides a magnetic field.

Work (1) Forcing a current through a resistance. (2) The product of a force.

Yield Strength The highest stress a material can stand without permanent deformation or damage, expressed in pounds per square inch (psi).

Yoke Sleeve Kit This can be installed instead of completely replacing the yoke. The sleeve is of heavy walled construction with a hardened steel surface having an outside diameter that is the same as the original yoke diameter.

Zener Diode A variation of the diode, this device functions like a standard diode until a certain voltage is reached. When the voltage level reaches this point, the zener diode will allow current to flow in the reverse direction. Zener diodes are often used in electronic voltage regulators.